Springer Series in Information Sciences 33

Editor: Thomas S. Huang

Springer-Verlag Berlin Heidelberg GmbH

Springer Series in Information Sciences

Editors: Thomas S. Huang   Teuvo Kohonen   Manfred R. Schroeder
Managing Editor: H. K. V. Lotsch

Thierry Viéville

A Few Steps
Towards
3D Active Vision

With 58 Figures

 Springer

Professor Thierry Viéville

INRIA, Projet Robotvis
Unité de Recherche Sophia Antipolis
2004, route des Lucioles, BP. 93
F-06902 Sophia Antipolis Cedex, France

Series Editors:

Professor Thomas S. Huang
Department of Electrical Engineering and Coordinated Science Laboratory,
University of Illinois, Urbana, IL 61801, USA

Professor Teuvo Kohonen
Helsinki University of Technology, Neural Networks Research Centre,
Rakentajanaukio 2C, FIN-02150 Espoo, Finland

Professor Dr. Manfred R. Schroeder
Drittes Physikalisches Institut, Universität Göttingen, Bürgerstrasse 42-44,
D-37073 Göttingen, Germany

Managing Editor:

Dr.-Ing. Helmut K. V. Lotsch
Springer-Verlag, Tiergartenstrasse 17,
D-69121 Heidelberg, Germany

Library of Congress Cataloging-in-Publication Data

Viéville, Thierry, 1959–
A few steps towards 3D active vision / Thierry Viéville. p. cm. – (Springer series in information sciences; 33)
Includes bibliographical references and index.

1. Robot vision. 2. Computer vision. I. Title. II. Series.
TJ211.3.V54   1997   629.8'92637–dc21   97-19673   CIP
ISBN 978-3-642-64580-8   ISBN 978-3-642-60842-1 (eBook)
DOI 10.1007/978-3-642-60842-1

Typesetting: Camera ready by the author using a Springer TeX macro package
Cover design: *design & production* GmbH, Heidelberg

SPIN: 10507842   54/3144 - 5 4 3 2 1 0 - Printed on acid-free paper

To Alice and Adèle

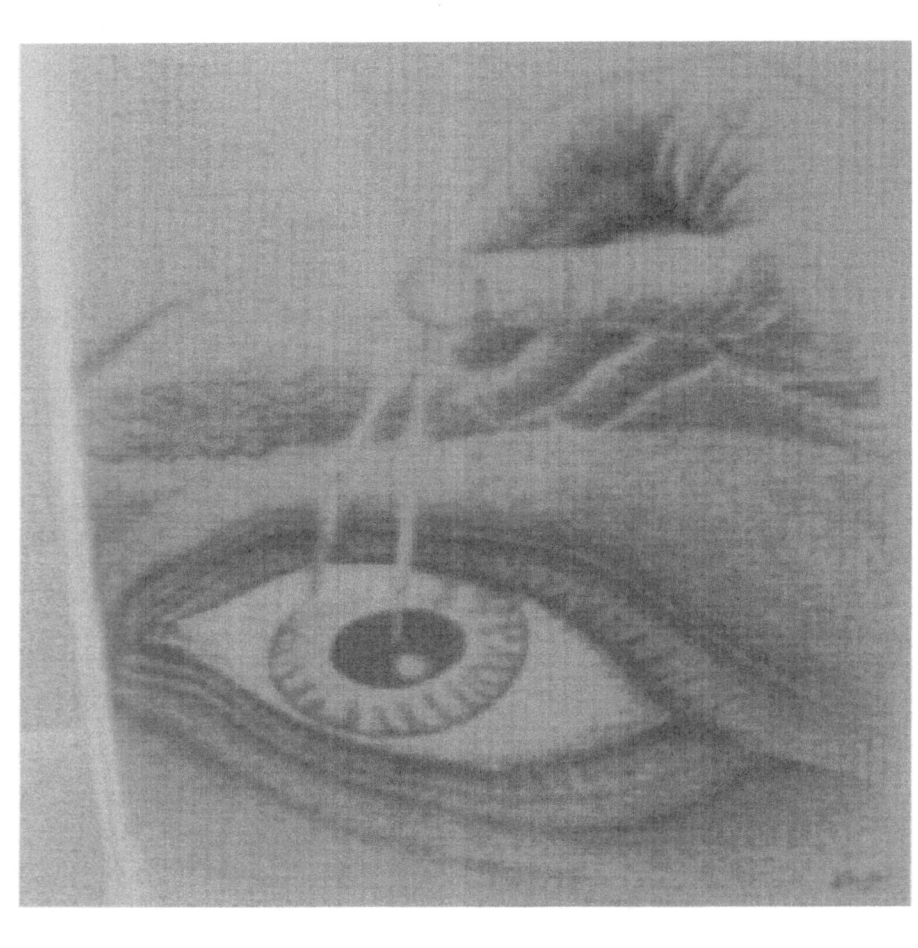

Preface

This book aims to analyse a specific problem in the field of active vision: how suitable is it to explicitly use 3D visual cues in an reactive visual task? In order to answer this question, we have collected a set of studies on this subject and have used these experimental and theoretical developments to propose a synthetic view on this problem, completed by some specific experimentations.

We first propose, through a short survey of the active vision problem, a precise definition of "3D active vision" and analyse, from a general point of view, the basic requirements for a reactive system to integrate 3D visual cues, and on the reserve, how can a reactive paradigm help to obtain 3D visual information. This discussion is intentionally very general and does not make any assumption about the particular architecture of the visual system, but is based on experimental data found in the literature, obtained on many other systems.

The next step is to verify whether these general ideas can easily be implemented, considering an effective active visual system. This is the goal of the second chapter in which we describe a mechanical system which has been designed to facilitate the fixation and tracking of 3D objects. The mechanics are described first and a framework to deal with eye-neck control is proposed. The visual mechanisms able to detect visual targets and effectively realise their 3D observations are described in the second part of this chapter. An important issue is that all these mechanisms are adaptive in the sense that strategies to automatically adapt their parameters are proposed.

A step further, the main problem when considering 3D vision on a reactive visual system is the problem of auto-calibration. The third chapter deals with this problem, in the case of a robotic head. The specificity of the approach is that deep involvements in difficult concepts of projective Geometry are avoided. Moreover, the auto-calibration strategy has been optimized considering a robotic head and the uncertainty about the calibration parameters is derived.

As soon as the mount is calibrated, we can – as in a biological visual system – make the visual layers of the system collaborate with other perceptual cues. The most relevant odometric cue, in this case, is inertial cue. This is studied in details in the fourth chapter, and shows how the usual structure of the motion paradigm can be simplified and enhanced when performed in cooperation with inertial cues. This also leads to high-level mechanisms of image stabilization, in which rotational and translational components of the system self-motion are treated separately.

As far as 3D visual perception is concerned, the retinal motion field is a very informative cue. However, we must consider that the camera calibration is not always known with precision, and that several objects with different motions are likely to be observed at the same time. This level of complexity is not covered by standard tools, but this gap is filled in the last chapter, where the motion field is analysed under these very general conditions, with robust mechanisms of estimation and several methods to segment object, in motion, detect planar structures, etc.

It is clear, at the end of these developments, that a reactive visual system, providing that it has been designed with some auto-calibration capabilities, is using inertial cues in cooperation with vision and is performing an analysis of the retinal motion in a suitable way, can definitely have 3D active visual perception capabilities. This has been demonstrated by both theoretical arguments and several experimentations. Does it mean that "the problem is solved"? Not at all. On the contrary, the gate is open now to realize much more complex reactive behaviors, based on 3D visual cues. This is briefly described in the conclusion.

Olivier Faugeras is gratefully acknowledged for some powerful ideas which are at the origin of this work.

A lot of thanks to Emmanuelle Clergue, Reyes Enciso, Bernard Giai-Checa, Hervé Mathieu, Luc Robert and Cyril Zeller which have highly contributed to some aspects of this work.

A big thank-you to Philippa Hook and Veit Schenk for their corrections.

Sophia-Antipolis, July 1997   *Thierry Viéville*

Contents

1. From 2D to 3D Active Vision

Let us start to analyse how suitable it is to explicitly use 3D visual cues in an reactive visual task. Let us firstly try to collect a set of studies on this subject and use these experimental and theoretical developments to propose a synthetic view on this problem, completed by some specific experimentations.

In this first chapter we propose, through a short survey of the active vision problem, a precise definition of "3D active vision" and analyse, from a general point of view, the basic requirements for a reactive system to integrate 3D visual cues, and on the reserve, how can a reactive paradigm help to obtain 3D visual information. This discussion is intentionally held very general and does not make any assumption about the particular architecture of the visual system, but is based on bibliographical experimental data obtained on many systems.

1.1 The Concept of Active Vision

Despite the large developments in machine vision taken in the past ten years, most research still involves small numbers of images being processed "off-line" with the resulting information being used to judge the success of an experiment, rather than to control another system such as a robot arm or autonomous vehicle or to realize a robotic or perceptual task. This is mainly due to a lack of knowledge of how to process image sequences efficiently and by the limitations of finite computing resources. However, this is not desirable especially if the requirements of many potential applications in advanced robotics are to be met. Steps are being taken to develop hardware capable of processing continuous image sequences and of delivering useful visual information such as depth maps etc., for the control of robot systems in real-time [1.1–17].

However, improvements to the processing hardware are by themselves insufficient. It has been known for some time that more flexible, dynamic vision systems are required for operation in uncontrolled, cluttered or hazardous environments. Improvements in the understanding of vision and computing power have added further impetus to this view with the growing realization that the simplifying assumption of static cameras may have made the vision problem more difficult. For example, it is clear that visual scenes are not uniform in the amount of information they contain. Hence an efficient vision system should apply variable amounts of processing (attention) to different locations. It is also clear that regions of high information/interest need to be tracked over time as objects move relative to the camera.

These ideas have been named "active-vision" and are now identified as a research topic on its own [1.18]. This research program is based on the following hypothesis:

For a robotic system, as for a biological system, the functionalities and the performances of the visual mechanisms are improved using reactive mechanisms controlling the visual sensor displacements and its internal parameters.

We usually distinguish:

(1) *Passive Vision* which is limited to the study of visual information. It corresponds to either:
    (a) one view (scene analysis),
    (b) two or three views at different locations in space (stereo),
    (c) two or three views at different times (motion analysis).
(2) *Dynamic Vision* which corresponds to the study of visual information, in an unbounded sequence of views.
(3) *Active Vision* in which there exists a feedback of the visual information on:
    (a) the visual sensor location, orientation and motion,
    (b) the lens and the image acquisition system (zoom, focus, iris, lighting),
    (c) the internal metric and the internal state of the system (visual calibration, early-vision smoothing, contrast thresholds).

Therefore active vision subsume dynamic and passive vision, but is somehow less general, since it must be adapted, or specialized, to a certain kind of system. This restriction will however allow it to be more efficient than applying directly very general principles.

From the control theory standpoint, the problems occurring are the use of vision data as feedback inputs in a robot closed loop control scheme. This concerns theoretical aspects like analysis and synthesis of the control laws as well as practical ones like the design of dedicated image processing hardware allowing us to implement these control laws in real time.

Furthermore, classical vision problems (such as 3D scene reconstruction and 3D object motion estimation) are to be reinvestigated with the active vision paradigm. The basic idea is that applying specific motions on a mobile camera can lead to new and robust solutions in several computer vision applications.

1.1.1 Can Reactive Vision Be "Better" Than Passive Vision?

The idea that vision algorithms can be improved by performing active visuomotor tasks with an adaptive and mobile visual sensor is well accepted but far from being obvious. On the contrary, the increase in system complexity might be a source of degradation in terms of performance, especially for 3D vision modules, when we want to address the problem of having such an active vision system dealing with high-level representations of the scene, especially considering three-dimensional (3D) representations.[1] Most of the earlier or recent studies in the field of active vision are based on mechanisms involving only bi-dimensional representations, that is representations in which three-dimensional parameters are implicit [1.18–25]. Obviously the relationship between the target 3D depth and the active visual system has already been introduced for the control of vergence, focus or zoom [1.23, 26, 27] but this "3D representation" is over-simplified since these models consider one punctual target in the scene only, except [1.28]. More recently, for an eye–arm system, the control of the three-dimensional motion of the camera has been related to the 3D structure of a rather complex object under observation [1.29, 30], but this result has not yet been extended to the case of robotic heads. The use of active head movements to help solving stereo correspondences has also being recently considered [1.31] but the precision of the mechanics for such a paradigm to be explored is, according to Luong and Faugeras, far from what is obtained with actual realizations [1.20, 23, 32, 33] and the only system with such a very high precision is yet only at a development stage [1.34]. So, paradoxically, if we want to output usable results concerning 3D active vision we first have to deal with monocular cues. At the computational level this does not mean considering the rigid displacement between the two images to be perfectly known as in the case of Stereo paradigms, but estimate it as in the case of Motion paradigms.[2] The facility in a motion paradigm is that we can assume the disparity between two frames to be small, leading to easy solutions for the correspondence problem. These correspondences must however be established (token-tracking). In addition to these difficulties, a crucial, mandatory and very sensitive problem is calibration. Both calibration of the robotic head and calibration of the visual sensor. Now, in the case of active vision, the extrinsic parameters and the intrinsic parameters of the visual sensor are modified dynamically. For instance, when tuning the zoom and focus of a lens, these parameters are modified and must be recomputed. It was thus a new challenge to determine dynamic calibration parameters by a simple observation of an unknown stationary scene, when performing a rigid motion as studied by [1.35–38]. The other alternatives were either to limit the visual analysis to some invariants [1.39, 40] or image based parameters not requiring calibration parameters [1.23, 41].

---

[1] We make the distinction between visual processes which deal either with internal representation involving only 2D parameters (3D parameters being implicit) or internal representation involving also 3D parameters (Camera calibration, 3D location, 3D orientation and 3D motion) explicitly.

[2] In such a framework, since 3D information is available for each camera, the Stereo problem is to be attacked as a "sensor fusion problem".

1.1.2 A Blink at the State of the Art in Active Vision

Considering the state of the art in this field, it was already clear that "active vision" does not simply mean driving the pan and tilt of a turret! More precisely, in Computer Vision, Active Vision is understood as both reactive (system tuning) and purposive (or task oriented) vision. The state of the art at the beginning of the project had been reviewed in the technical annex but we must now analyse the evolution of this area during the last three years.

Recently, a part of the American and international vision community, during a workshop held at Chicago, has attempted to define in rather great detail what is active vision and which topics are covered by this field of research [1.18]. Active vision has been defined as *active control of all camera parameters*. The basic ideas were to attempt to simplify early-vision problems, allow a selection of the visual data to reduce the inherent problems encountered in vision (too much data, most of them being redundant, information being episodic (clumped) and difficult to extract) and finally expand the capabilities of the vision systems. Several aspects have been clearly identified:

(1) *Attention processes.* The goal is to selectively reduce the data to be processed. Selection being done by location (region of interest), motion, depth, data compression or effective early vision.
(2) *Foveal sensing.* This area is concerned with the development of specific visual sensors (variable resolution sensors, biologic-like retinas, etc.) with, for instance, high-resolution at the center and lower resolution at the periphery.
(3) *Gaze control.* This means *control of the camera parameters and control of the visual processes to acquire images which are directly suited to the vision task.* This topic is subdivided into two areas:
    (a) *Gaze stabilization*indeximage/gaze stabilization, i.e. maintain clear images of some world location, and reduce the disparity between two frames.
    (b) *Gaze change*, i.e. changing parameters to meet new requirements of the perceptual task. This problem is also referred as *"where to look next".* In fact, this area is poorly understood, even if some classifications (task driven, feature driven, environment driven) have been proposed.
(4) *Head–eye coordination.* This refers to the use of vision to predict the motion of a robotic sub-system, while data from this sub-system is used backwards to correct the visual perception. This includes manipulation of objects to aid vision analysis, auto-calibration, visually guided assembly tasks, etc.

In addition to these four topics, requirements in three areas have been identified by the Chicago group: (a) mechanical hardware, (b) electrical hardware, and (c) benchmarks. They also have quoted several potential applications such as autonomous navigation of vehicles, industrial tracking and surveillance.

1.2 A Short Review of Existing Active Visual Systems

1.2.1 Active Visual Sensors

The simplest and most common stereo vision systems use a pair of cameras in fixed mountings with their optic axes parallel [1.42]. The geometry of such systems makes depth recovery straightforward, and the epipolar lines can be arranged parallel to the image rasters so that the images do not need registering and so that matching corresponding features is easy. However, such a system is unrealistic and far too restrictive for real applications, so the most recent stereo work began by considering two or three cameras capable of being set up with non-zero vergence and, occasionally, non-zero gaze angles [1.43–47]. Completely general rigid rotations and translations between one camera's frame and another were first considered in photometry [1.48], but such an arrangement has now become the norm in computer vision [1.49–52]. General systems require careful a calibration of this transformation and of the interior geometrical parameters of the cameras [1.53–56]. However, these computational disadvantages are outweighed by the system generality and the greatly improved working volume in which objects may be simultaneously imaged by a set of cameras with vergence.

In [1.20, 57], the camera platform had four degrees of freedom – vertical motion, horizontal motion, pan and tilt – while the two (Charged Coupled Device) CCD cameras used could verge symmetrically and adjust their focus, aperture and zoom independently. The lighting was also controlled. The system was built from low cost equipment and many components were not engineered precisely. As a consequence, it was not used for accurate stereo vision and was not as accurately calibrated as fixed stereo systems. It was used to provide very coarse, approximate stereo and depth from focus measurements at a small number of points (approx. 10–20 per view) which, if consistent, were merged by simple averaging. Despite the small number of features in each view, no attempt at real-time operation was made as many of the calculations were carried out on a time-shared computer and the aperture and zoom (depth of field, and field of view and magnification) conditions required for the stereo and depth-from-focus conflicted. The two modes were used co-operatively, however, to cross-check the depths inferred from each other.

In the second system, Cornog [1.19] used an electromechanical two-camera positioning system (the eye–head robot) to develop and implement control algorithms for the guidance of "eye" and "head" movements during the tasks of fixating and tracking an object in 3D. The emphasis in this system was on the control loops and on a comparison of the robot system with primate oculomotor systems. The robot apparatus used was designed in 1982 from preliminary ideas and specifications developed by the researchers [1.19]. It consisted of a moving platform (two axes of rotation, pan and pitch, driven by stepping motors) upon which were mounted two solid state cameras and four rotatable mirrors deflected by galvanometers. The mirrors provided rapid "horizontal" and "vertical" deflection of the line of sight of each camera, independently of the pan and pitch of the platform and of movement of the other camera's mirrors. The system thus had six degrees of freedom, more than enough for tracking an object in 3D which only requires three. Since the platform moved both cameras and all four mirrors together, only the mirrors could be used to change the vergence of the lines of sight (range), but both the platform and the mirrors could be used to change the elevation and gaze. Deciding which to move was therefore a question of the "eye–head" control strategy in which the mirrors were used to track quick motions of small amplitude while the stepping motors were used for slower reorientations over a larger range.

This system was used to track a high-contrast target, a white sphere placed against a dark background, that could be easily and quickly located by a threshold mechanism. The tracking algorithms, implemented in the LISP computer language, allowed the eye–head robot to track the target moving up to 15 deg/s for brief periods and to hold the image stable to within 3 pixels. The tracking combined velocity matching (to stabilize the images of the target) and fixation (to keep the images centered on the cameras), the integration of which required: (a) the addition of a predicted error due to target velocity to the positional error in order to calculate the amplitude of mirror movement; and, (b) blanking of the velocity calculations during the reorientation. The strategy adopted for control of the stepping motors upon which the "head" platform was mounted, was to keep the mirrors roughly centered in their rather limited range of deflection.

The vision in this system left much to be desired as it utilized only crude, primitive processing tailored to the characteristics of the target object. In addition, targets could only be tracked for brief periods because of the low speed (4 deg/s) obtainable from the platform on which the cameras were mounted.

The most successful attempt to construct a gaze control system has been carried out at the University of Rochester. The Rochester Robot (RR) [1.32] consists of two CCD cameras mounted on a robot arm. The cameras can rotate independently in the plane of the platform, while the 6 degrees of freedom arm on which the platform is mounted is only allowed to rotate in the vertical plane. Camera rotations are faster than arm movements as they are driven by stepper motors with a peak velocity of 400 deg/s, a mean velocity of 60 deg/s and a resolution of 0.14 deg. A number of the control loops involved in gaze control have been implemented and shown to work in real time (pursuit and vergence) using a pipeline parallel image processing system that carries out low and intermediate level image processing at 30 frames/s [1.26]. Most recently, simulation studies have been carried out on the problem of combining these control loops [1.21]. Although the RR is more advanced than the previous attempts it suffers from one of their problems – that of having to use inappropriate equipment adapted from other sources. For example the interface of this robot involves a number of levels and injects a long and variable delay into the control system.

Gaze control is thus essential for machine vision and the demonstration by the RR of its plausibility have spawned a number of other gaze projects in the United States at: University of Illinois [1.58], Harvard University [1.59] and Boston University.

In Europe, following this pioneering work in the USA, there is considerable interest in moving camera systems at a number of British, French and Scandinavian academic institutions. In particular, two projects are now having such systems under development: on one hand, the Royal Institute of Technology, in Stockholm, developed a new very powerful binocular system with the capability of controlling not only the optical axis orientation but also different parameters of the camera locations such as the baseline, the rotation center, etc.); on another hand, the European project named Real Time Gaze Control) RTGC was developing a low cost fast system of binocular camera, and in the future.the previous formalism will be experimented on this system.

1.2.2 Control for Active Vision

Current approaches in designing sensory-motor tasks might be divided into three classes. In the first class researchers have tried to formalize sensing behaviors as a very general optimization problem, and defined the objective of the sensing behavior as a criterion to minimize. They implement the behaviors using minimization techniques [1.60], including optimal control techniques [1.61]. In a second class scientists use the concept of "task-function" as defined by [1.62]. This approach has already yielded promising results [1.63, 64] in vision, and several slow real-time experiments have been realized. It is however much more suitable to low-level control loops using the visual sensor as an input, than to design complex sensing behaviors [1.65]. The last class of studies, including neuro-physiological model of the oculomotor system, define sensing behaviors as a combination of "black boxes", building a functional model of the system [1.66]. There is however a lack of studies on the way to combine different techniques at different levels of speed or complexity. Despite the existence of relevant studies on integration of sensing behaviors, and perception system architecture [1.67, 68], there is no result about the stability of complex behaviors (except for [1.62]).

In order to fill this gap, a specific architecture of behaviors have been proposed [1.33]. Such "oculomotor" behaviors have to be used in cooperation with vision when intending to realize an active perception of the visual surroundings. The specificity of this approach was to make use of a powerful algorithm in control theory [1.69]. An optimal estimator and controller, for a sensory-motor linear system has been described. Then a specific architecture to deal with real time implementations but without giving up the use of nonlinear constraints and the control of the stability, have been proposed.

It is now clear that during a tracking task or during the visual capture of a target, acquiring 3D information on the object of interest allows to (a) easily predict the target location and motion, and thus – using some dedicated technique [[1.69, 70] – it is possible to compensate, at the level of a visual control loop, for the delays introduced by the processing. Moreover, in the case of target capture, (b) the knowledge of the robotic system calibration parameters allows to compute directly the final configuration to attain, without the use of iterative technique. When 3D information is available the system not only controls the gaze direction, but also vergence, focus and zoom. In addition to that, (c) the evaluation of the 3D parameters is performed using several cues obtained because we are on an active visual system (focus, vergence and motion cues). On a restrained hardware, such as [1.71], 3D visual loops can take place as simple look-and-move paradigms. Finally, as reported in [1.30, 71], reactive 3D vision not only allows to track or capture a target but can also be used to actively perceive a 3D visual object, and derive optimal displacements in a structure from motion paradigm.

The problem of tracking other visual tokens such as snakes [1.72] but with the goal of recovering the 3D parameters of the curve are reported in [1.73]. Real-time implementations of these mechanisms have been made [1.74].

Finally a negative but very interesting result has been issued by [1.31]. Can active head displacements help solve stereo correspondence? The answer is no. More precisely, considering a realistic configuration the answer is: in theory yes, and quite easy, but in practice the required mechanical precision is unacceptable.

1.2.3 Auto-Calibration of an Active Visual System

The use of zoom and auto-focus modifies the intrinsic calibration parameters of the camera and the parameters of the camera positioning system. In addition, the use of variable focal lengths (especially small ones) introduces geometrical distortion for objects which are not close to the optical axis. It is therefore mandatory to develop methods to automatically recalibrate the visual sensor [1.75].

The goal of this research is thus to develop robust, on line, calibration techniques for keeping an accurate model of the intrinsic and extrinsic camera parameters. The calibration procedure must be run continuously since all those parameters are changing continuously, must be capable of operating directly on the environment without the use of artificial patterns, and must be capable of integrating high level data.

Closely related to this problem of "intrinsic calibration" is the problem of recovery of parameters of the camera positioning system and of camera location with respect to the visual surroundings ("extrinsic calibration"). On the other hand, knowledge about camera location/motion might be used by the auto-calibration algorithm [1.38].

Conventional calibration techniques are based on the observation of already known patterns, but here calibration has to be performed "on the environment". Since this a fairly open subject, the first developments were purely theoretical speculations and lead to a very deep analysis of the geometrical constraints related to the auto-calibration problem [1.76]. The problem has been solved in a completely general situation [1.36, 77] including some degenerated cases [1.78].

As a fall-back position, recalibration has also been roughly performed either using off-line static methods [1.71] or in some specific situations: (1) when it is possible to detect points at the horizon such as in [1.79]; in this situation the auto-calibration problem is better defined and can be solved efficiently [1.71]; (2) when it is possible to rotate around a fixed axis, the relative angles of the rotation being known [1.38, 80], in this situation, auto-calibration can be performed in a more robust way, the uncertainties related to the intrinsic and extrinsic parameters can be estimated, and some qualitative information about the optimal displacement to performed are issued. These methods are to be applied in situations were the original method is either unusable (singularity) or not very efficient. These two additional contributions are thus a complement of the general method.

1.2.4 Perception of 3D Parameters Using Active Vision

Segmentation of objects, estimation of their motion and location is best done in 3D, since their real motion and structure is much simpler than their projection in 2D. This is also what is needed for applications such as surveillance or autonomous navigation. Its interest for active vision is to provide a quantitative measurement of the surrounding objects, to allow a 3D reconstruction of the visual environment, and to provide the basic cues to recognize moving objects, obstacles, etc.

For instance, a module of incremental self-motion and 3D structure recovery to the case where: (1) an estimate of the vertical is available and (2) the visual tokens are defined using qualitative information about their 3D parameters (3) active visual motion are performed [1.79, 81, 82]. This algorithm corresponds to an original functional architecture for a 3D vision system. It has been designed as a compromise between fast implementations and robust computations.

It has been also demonstrated that, in an active visual system, odometric cues effectively help recovering the 3D structure of the scene [1.83]. Moreover, the fusion between inertial cues and 3D vision, as originally planned in the project yields very powerful reactive mechanism as for instance in autonomous navigation [1.84].

Real-time dynamic map elaboration from token-trackers technic has also been studied. The sparse approach for the computation of visual motion can lead to fast implementations of dense parametric maps of motion fields with detection of unexpected motion in this map [1.85], the method being as efficient as a technique based on the fusion of motion and stereo [1.86].

In addition, re-implementations of the early-vision mechanisms, compatible with an active visual loop, to be partially implemented on the actual hardware, or to be implemented in other architectures have been achieved [1.87, 88].

Finally, there is a relatively simple answer to the problem of "where to look next". Considering attention focusing we can very easily give a list of which objects in a scene are to be normally observed by an active visual system: (1) Moving objects might trigger potential alarms ("prey", "predator", moving obstacles to avoid). (2) Nearby objects will be the first to interact with the system as potential obstacles. Their proximity should ease 3D observation. (3) Objects with a high density of edges correspond to informative parts of the visual field and might be worth a closer look. These three categories might appear as very natural, indeed they are, but in addition, they correspond to a very precise 3D motion property. In fact, they can be identified in a scene as soon as the rotational disparity has been canceled between two consecutive frames [1.71, 82], which is going to be developed in the next chapter.

1.3 Architecture of an Active Visual System

1.3.1 The Three Main Functions of an Active Visual System

We know that using appropriate camera movements a visual controlled system can stabilize the image on the sensor retina [1.82], contribute to visual exploration of the surroundings [1.28], or track an object in movement [1.89], but it can also provide a certain estimation of the self-motion, and of the structure and motion of certain parts of the visual neighborhood [1.23, 30].

As a robotic system, the realization will have to deal with a certain set of tasks. These tasks are not be done alone by the control system of the sensor, but only in coordination with higher level of intelligence.

We can identify the following tasks which are easily related to what appears the human oculomotor system [1.90]:

Perform, as Quickly and as Accurately as Possible a Certain Movement with the Visual Sensor. Higher parts of the system might require, for a given observation, to direct the visual sensor towards a certain object. This task might be designed as:

– A geometrical transformation allowing to optimise the relative position of the sensor with respect to the observed object.
– A generation of the sensor trajectory, in order to realized the movement.

This process takes a positioning requirement as an input, and provides the position of the sensor as an output, with some additional information if the movement is not possible (obstacles, end of the system stoke). It internally generates saccadic movements of the sensor.

The positioning requirement of the visual sensor might be expressed:

- With respect to the robot; in this case the geometry of the mechanism which moves the visual sensor has to be known.
- With respect to the stationary surroundings; the self-motion of the robot and its orientation in space has also to be integrated.
- With respect to an observed object; the relative position of the object with respect to the sensor has to be known.

In the next task, all three frames of reference have to be used. Since they have to be computed by the control system anyway, we propose to have them available in this process.

Stabilize the Sensor in Order to Obtain Good Data During Acquisition. Visual observation is blurred when relative motion occurs between the camera and the observed object. Then a mobile robot cannot have vision if the camera is not stabilized. This task might be understood as follows:

- While it is possible, fix the camera on the part of the visual surrounding which is to be observed (by robot movement compensation and observed object tracking).
- Detect if the observation is no more possible (the movements are too important, or the stabilization process has drifted and a movement of reset is necessary), and stop the acquisition for the time a rapid recentering movement is performed.

This process takes as an input the representation of the part of the visual surroundings to be observed and provide as an output an indication whether or not the visual observation is possible. Internally it controls visual sensors movements as a function of its internal representation of the motions.

In order to be accomplished the process has then to:

- measure robot movements and compensate them,
- identify the movements given to the visual sensor to adapt them,
- estimate the observed objet movement to track it.

It can also provide this information to higher layers of the robot, as detailed in the following.

Provide a Permanent Estimation of Motions and Relative Positions. Movements will be first measured from inertial and odometric sensors. Precisely, the following information is to be provided:

- Absolute orientation of the robot head with respect to the vertical (measure of the gravity orientation),

– Orientation of the robot head in the horizontal plane (gyroscopic or odometric estimation),
– 3D Angular velocity of the robot head (gyrometers based on Coriolis
forces),
– 3D Linear acceleration of the robot head (linear accelerometers with elimination of the gravity acceleration),
– Relative position and orientation of the visual sensor with respect to the
robot (odometric information).
– Angular position and velocity of each join of the mechanism which moves
the visual sensor (odometric information).

The visual surrounding of the system is composed of rigid objects, not
necessaily piece-wise planar, eventually partially occluded, and is divided in
three categories:

– a stationary background,
– one object, in motion or not, under observation.
– some moving object one does not take care of,

Then, this module must provide a sorting of the visual tokens in these
three categories. This process takes a certain set of visual tokens as an input, and returns different estimations of motion and position, with a primary
analysis of the picture as an output. All this information has to be provided
with an estimation of the error of computation performed on each parameter.

In fact, these functions will not necessarily be implemented separately
but will be integrated in some comprehensive mechanism able to automatically switch from one mode to another.

1.3.2 Basic Requirements of a Visual System

In this context, the system has to be designed to work with a continuous
flow of visual information, a sequence of images, and not only a few sets of
frames. Moreover self-motion and camera-motion estimates are provided and
must be integrated by the system. This information corresponds to mount
positioning[3] and to inertial cues [1.82, 91].

It is of primary importance to deal with simple tokens, since we want to
have a continuous analysis of the visual flow of information and be able to
limit the amount of computation required. However, we must not rely on a
sparse description of the image, because – considering that we want to detect
any potential alarm, for instance – we need to have a dense description of the
information in the image. Hopefully, we can expect small movements from

---

[3] In the case of a robotic head or an eye–arm system, we call "mount" all the
mechanical assembly of camera(s), controllable lenses, and rotational or translational degrees of freedom.

frame to frame and drastically simplify the token matching problem. This will be used in our implementation.

We thus want to:

− reduce as much as possible the amount of information,
− use noiseless primitives with respect to motion parameters,
− take odometric cues into account,
− simplify the matching problem between two frames,
− perform computations in real time.

Then, visual information is treated at different levels: By the control mechanism, low-level treatments are applied on the picture in order to extract some useful information about motion. By a higher layer of the system, which can use the preprocessed information.

A set of computation modules is identified: *high-rate data flow treatments, with an internal clock of about* 1 kHz, *which are*: the real-time control of the eye-movement effectors, and trajectory generation, the real-time coordinates transformation from eye-movement effector joins to Cartesian coordinates, the real-time analysis of inertial information, including separation between rectilinear accelerations and gravity, the interface with higher layers of analysis; *video-rate pretreatments, with a picture acquisition of* 25 Hz, *which are*: video picture acquisition, real-time convolution of the picture with four operators: picture filtering, horizontal and vertical gradient, maximum contrast pixels detection, which will be the "picture primitives"; *low-rate treatments of the picture primitives* 5–25 Hz, *which are*: real-time extraction of picture tokens or primitives (edges, corners, points), primitives matching between two frames, and sorting in different rigid objects, computation of visual-motion parameters; *calibration modules: calibration of:* camera intrinsic parameters, inertial sensors parameters, mechanical joints parameters.

1.4 2D Versus 3D Vision in an Active Visual System

1.4.1 Active Vision and 2D Visual Servoing

Many visual tasks such as target tracking, image stabilization, motion detection, beacon detection, object recognition, etc. do not require 3D algorithms whereas 2D (or image based) algorithms are sufficient and often more robust and efficient. In such frameworks [1.23–25, 30], the 3D parameters are only implicit.

Recent advances in vision sensor technology and vision processing now authorize the use of vision data to control real time robotic systems. This approach is called "visual servoing" or "vision based control" [1.22, 62, 64, 70].

It is particularly well adapted when the task can be expressed in a natural w·) as a control of the relative attitude (position and orientation) of the ca.,nera with regard to a particular part of the scene.

In a visual servoing approach, the basic idea consists to assume that a task is perfectly achieved when a desired image of the scene corresponding to a particular viewpoint of the camera, has been reached. In terms of control, this can be stated as a problem of regulation of a certain output function directly defined in the image frame. Implementing a visual servoing application needs to solve several problems.

As a first step, we have to specify the task in terms of signals to be controlled in the image which constituted the visual feedback. The second step requires the choice of the image processing methods for extracting these signals. The last step consists of computing a robust closed loop control scheme based on the vision data which allows from an initial image seen by the camera to converge to the desired one [1.62].

The most simple task we can consider is to control the camera orientation such that the image of an object appears at a paticular locations in the image: for example in the different corners or in the middle of the image. The two degrees of freedom controlled to perform this task are here the camera pan and tilt. Camera self-motions can be combined simultaneously: for example, translational motion along the optical axis (zoom), rotational motion around the optical axis or, more generally, motions using all the degrees of freedom of the robot non constrained by the vision-based task. For the case where two different tasks are considered (a vision-based one and a trajectory tracking), the redundancy framework of the task function approach allows to perfectly combine them [1.63].

Since in visual servoing, camera motions are computed with only bi-dimensional data extracted from the images, the approach is exactly the same when the considered objects are mobile in the scene. Meanwhile, an estimation of the target motion in the image has to be introduced in the control law in order to compensate this 3D motion which generally implies tracking errors. This estimation can be obtained using classical control techniques such as integrator loops or with the Kalman filtering approach (with correlated noise in the presented case). Work is done in order to detect and compensate target maneuvering (which appear when abrupt changes in the nature or in the amplitude of the object motion is observed) and to estimate the tri-dimensional object motions using control data needed for the tracking.

Moreover, all visual processes are indeed measuring data in the images and, since we are dealing with monocular cues, the measures are always 2D primitives (such as points, line-segments, curves). At this stage everything is "2D". Specifically, in the case of real-time gaze control, it is *a priori* a better choice to design several 2D fast and reliable visual mechanisms [1.23] than to attempt to elaborate heavy, complex, slow, and fragile 3D based visual loops.

1.4.2 Introduction of 3D Vision in Visual Loops

Hence, when introducing 3D parameters in a visual representation, we must be aware that, as usually when increasing the dimension of the state of a system, computation time will increase, system observability and stability might be affected and the estimation is likely to be less robust. This explains why it is often desirable to "work in 2D".

But on the other hand, can we really perceive our visual environment and perform all visual tasks without 3D? Obviously not. More precisely the right question is: *When is it mandatory to introduce 3D parameters in the internal representation of a visual system?* The answer is "quite often". It is very easy to list many actions in which 3D vision must be introduced: knowing the exact value of the focus, calibrating the vergence of a binocular system, calibrating the system visual metric to find the relationship between the angular position of the mount and the retinal displacement in pixels when a translation occurs, computing relative depth, evaluating the depth of an object, positioning with respect to a 3D map, etc. Related applications are: vehicle positioning and manoeuvres, object observation and recognition, moveable or stationary obstacle avoidance, 3D mapping for surveillance, etc. We thus must overcome the previous difficulties and also introduce 3D computations in an active visual system.

In fact, there are two ways to introduce 3D vision in a reactive visual system: either as a passive or as an active visual process.

Let us assume, for instance, that an object of interest is tracked using 2D vision only (i.e. tracking cancelling an error signal computed in the image). In parallel to this "fast tracking" the system can perform structure recovery from "known" motion, because the motion, or part of the mount motion, is known. This is due to the fact that motions executed by the mount are measurable. Moreover, because inertial cues can be available [1.82, 91], the system self-motion is also partially known and its spatial orientation is obtained in two directions (orientations with respect to the vertical). Such knowledge could be either quantitative (a value with a certain associated error) or qualitative (the direction or the amplitude of the rotational or translational nature of the motion is known). Because of this, the system is capable of improving the original "structure from motion" paradigm or perform it in a more general way (for instance recovering the calibration parameters) [1.38].

Additionally, if the system is not actually performing another visual task one can generate suitable motion of the mount to perform 3D scene analysis. In this case the dynamics of the motion or the geometry of the motion can be specified (it is known that certain kinds of motion will be more suitable for calibration [1.38] or for structure from motion estimation [1.29, 30]). The previous paradigm could be done with some optimality. Roughly speaking, we want to perform displacements of the mount which generate a "relevant" disparity between two frames.

1.4.3 Basic Modules for a 3D Active Visual System

If we want to perform the previous tasks, or similar ones, what are the "reactive mechanisms" required for such a system? Our analysis, based on previous works in the field [1.25, 33, 38, 82], yields four of them:

Early Vision and Lens Parameter Tuning. It is not possible during active vision to use an early vision system for which one must manually adjust the internal parameters (there are up to 16 such parameters in a real-time vision machine [1.33]). However, these parametric adjustments are mandatory because they allow the system to be adaptive and usable in varying conditions. Thus due to the photogrammetric variations expected during active vision, it is not possible to keep fixed average values for parameters such as lens iris, acquisition gain and offset, contrast thresholds, etc. It is important to have an adaptive system able to tune these parameters. The early-vision system must not require manual adjustment. This module is reactive since visual perception induces changes in lens, acquisition system, and early-vision hardware parameters. It is not a "3D mechanism" by itself but it is mandatory for 3D vision.

In almost all implementations, auto-focus, automatic adjustment of iris in synergy with gain and offset adjustment are utilised using low-level visual loops. Contrast thresholds are also automatically adjusted, using statistical models [1.71].

Auto-Calibration of the Visual System Intrinsic Parameters. If one changes any lens or mount parameters the visual calibration is to be recomputed [1.36, 38]. But if we consider a rigid object with enough tokens, satisfying enough rigidity constraints, it is possible to obtain extra equations during its motion involving only calibration parameters. Using these equations, calibration parameters can be recomputed whatever the relative motion of the observed object is (passive auto-calibration) [1.36]. Additionally if one can control mount displacements, it has been shown that performing an off-centered rotation around a fixed axis is particularly suitable for intrinsic parameters recovery (active auto-calibration) [1.38]. Similarly if a pure rotation around the camera optical center of the camera is induced auto-calibration is also computable [1.35].

In some implementations, because of the relative simplicity of the mechanical design, we can avoid using such sophisticated mechanisms but have *pre-calibrated* the mount in a large spectra of possible positions and obtain the extrinsic and intrinsic calibration parameters at any time, by interpolating the pre-recorded values.

Picture Stabilization Processes. If there is no disparity between two frames, there is no information about the projected motion, and no possibility to analyze dynamically the visual scene. Conversely, provided a model of structure and motion is available, one may cancel all the disparity between two consecutive frames by a suitable "reprojection". This means that we have extracted all the information about the projected motion, hence allowing perception of structure and motion. By such a mechanism, we obtain the scene parameters if and only if the measure is unambiguous. In other words, an image sequence is perfectly stabilized if and only if 3D knowledge allows to cancel the disparity between two frames ; and no ambiguity occurs.

In particular we can reproject, for each picture, the visual tokens in such a way that (1) the vertical in the image is aligned with some absolute vertical, and (2) the average disparity between two frames is minimum. This stabilization might be performed either using degrees of freedom of the mount, or by an internal reprojection or *rectification*. The vertical cue is obtained from visual and inertial cues, or odometric sensors [1.82].

Such a task is crucial for reactive vision. It first allows to reduce the ambiguities in the token-tracking problem, because the expected disparity between two occurrences of a token is reduced by this mechanism. It also simplifies the computation of structure and motion since it can be shown, if the system is calibrated, that one can cancel the rotational disparity between two frames, either using visual or inertial cues [1.82].

This mechanism acts either on the mount (control of gaze direction), or simply controls the internal metric of the system (picture reprojection). Moreover, since a rotation around the optical center corresponds to a particular homographic transform of the image [1.35, 40, 82], we can design "virtual" degrees of freedom, for instance eye torsion, and combine them with mechanical degrees of freedom for the control of image stabilization. In fact, using homographic transformations of the image plane for stabilization allows to work without calibration and compensates for complex projected motion, as discussed in [1.92].

A key idea in this book is that, because (1) the system is calibrated and (2) inertial sensors computing the angular velocity of the camera are available, the rotational disparity is automatically computed in a straightforward way, using odometric cues only. Therefore, picture stabilization is performed using odometric cues only.

Interaction Between Control of the Mount and 3D Perception. Obviously, it is possible to control zoom, focus and gaze direction including vergence using only 2D cues. Considering the projection of an object under observation, it is straightforward to tune the zoom such that the object size has the desired extent, adapt focus to minimize blur, and perform vergence (plus pan and tilt) to view the object around the retinas center in both cameras [1.23]. However, the previous method requires feedback control and different object characteristics must be recomputed at each step. Furthermore, if the depth and the size of the object are known and if the system is calibrated, these parameters can be estimated in one step. Conversely, the observation of vergence angle, of the focus yielding a minimum blur, and of object projection behavior during zoom, provide a coarse estimate of object size and depth thus usable if the system deals with 3D parameters.

One very important consequence of this "one-step" estimate is the following: control is not necessarily embedded in a real-time feedback, but could simply be limited to static positioning. In other words, the control can be a simple "look and move" paradigm. This restriction is very important in the case of hardware limitations. Thus, using 3D cues, direct global control is possible, the local feedbacks (implemented using 2D cues) being only used for residual error correction.

In almost every implementation, the 3D control of the mount is based on the internal computation of the **average depth** of 3D object under fixation. This simple representation allows to relate several active visual cues, and simplifies the control of the mount.

1.5 Gaze Control in 3D Active Visual Systems

Let us finally explain in detail some requirements when controlling the mount parameters of a robotic head on which 3D vision is to be performed.

Within the present framework, we consider a "robotic head" or "mount" made of a "stand", a "neck" performing off-centered rotations, and an "eye" performing rotations around the optical center.[4] Instantiations of such a mechanism are now very common [1.20, 21, 23, 33, 34]. However, whereas almost all algorithms in the field deal with very simple targets [1.19–21, 23–25, 29, 30, 89] we would like to design a more sophisticated behavior, but we must not only modify the visual algorithms but also the mechanical hardware. Let us explain why.

[$\mathcal{R}_1$] We need to reduce (but not necessary cancel) the disparity between two consecutive frames (retinal stabilization) because: (1) Vision algorithms for motion estimation (token-trackers, tracking-snakes, optical-flow operators) perform better when the disparity between two consecutive frames is minimum [1.82]. (2) Calibration parameters are valid only for a given area of the visual field thus only if the observed object has a stable position on the retina. (3) This will maintain the observation of an object during several frames as required by 3D vision algorithms.

[$\mathcal{R}_2$] As in nature, it is convenient to maintain the observed object close to the estimated principal point (foveation) since: (1) The calibration model relies on the pin-hole model which is valid only close to the optical axis (a biological equivalent is the fovea). (2) A moving object with unpredicible motion is easily maintained in the visual field if it stays near the center of the

---

[4] It has been established [1.23, 34] that one can design systems which really perform pure rotations with respect to the optical center of the camera (more precisely the object nodal point); the related errors are negligible for objects with a reasonable depth. These degrees of freedom often include vergence.

retina. (3) The induced disparity remains minimal when zooming (see $[\mathcal{R}_1]$), if we are close to the principal point.

Obviously, from a functional point of view, $[\mathcal{R}_2] \Rightarrow [\mathcal{R}_1]$, i.e. if the observed object is maintained close to the retinal fovea, it is indeed stabilized. Generally, the first requirement is related to a "velocity" since the object retinal drift is to be cancelled, whereas the second requirement is defined as a constraint in position. Therefore, the implementation of the two mechanisms are not necessary equivalent, but can be complementary as in the human oculomotor system.

$[\mathcal{R}_3]$ When $[\mathcal{R}_2]$ is not satisfied, to prevent observed objects from drifting out of sight we must reset the gaze direction as quickly as possible. This can be done by a known rotational displacement (saccade) because: (1) An off-centered object will drift rapidly due to its eccentricity. (2) The system is not likely to manage high disparities for such rapid drifts.

A reasonable strategy is to reset the mount position as quickly as possible, cancel visual information during the motion, and restart visual perception when it ends. However, we must relate the visual information before and after the displacement. If this displacement is done with a translation, image deformations will occur and the correspondence between the two images is not straightforward (it depends on the depth of objects). On the other hand, if the displacement is simply a rotation, the related image transformation is a simple and predictable reprojection, hence the system can relate the visual information before and after displacement. Furthermore, if this saccade is very quick, the additional variations in the image due to non-stationary objects between the beginning and the end of the motion are expected to be small. Therefore, it is a good strategy to induce relatively fast rotational displacements of the eye.

$[\mathcal{R}_4]$ Translational motion of the mount is also required to infer structure from motion and to relate the 3D structure parameters to the projected displacement. This can be done with linear joints performing pure translations or off-centered rotations with angular joints. The orientation of the translation must be "orthogonal to the 2D points" [1.29, 82]. The instantaneous amplitude of the translation is not constrained. The global observation can be performed combining several displacements into a desirable high-amplitude motion. This yields the following mechanism: *induce a translational motion in a direction orthogonal to the average target location in the image, and compensate for this disparity using rotational stabilization (eye–neck coordination)* [1.25, 82].

When the robotic head is made of a "neck" (off-centered rotations) and an eye (pure rotations), the neck displacements are used to limit the angular displacements of the eye in its "orbit". One very simple, but efficient strategy [1.23, 25] is to require the neck position to take the average position of the gaze over a certain time-window. Equivalently, the neck displacements compensate for the slow gaze motions, and the eye displacements for the rapid gaze motions. In [1.25] the time window is adjusted to be compatible with neck dynamic, but avoid the eye to move out of its displacement bounds. This strategy is compatible with $[\mathcal{R}_3]$.

$[\mathcal{R}_5]$ Zoom control, when used to cope with variations in disparity, is subject to contradictory requirements because: (1) To detect unexpected objects, the best configuration for the zoom is to be minimum (the focal length being smallest, the field of view is wider). This extremal configuration corresponds also to a situation where object size and projected displacements are minimal as required from $[\mathcal{R}_1]$. (2) If the field of view is too wide then so will be the density of edges and an artificial disparity will be induced by matching errors. Zooming into the observed object will overcome insufficient resolution. This leads to a criterion for zoom control: *the focal length is to be increased if and only if this reduces the residual disparity between two frames for the observed object, and it is to be tuned to minimize this disparity.* In some case this should be done in a "saccadic mode" because most vision modules assume the projection matrix to be constant [1.25, 36] when the system is to be dynamically calibrated. This is not a restriction for our implementation, since through calibration is always given.

One way to avoid this contradiction and preserve both requirements 1 and 2 is to use to *unhomogeneous* visual sensor. One with a large field of view, the other with zooming capabilities. We have implemented this strategy, and will show that stereoscopic perception is still possible with a "peripheral" and a "foveal" sensor.

Using the previous discussion we can easily integrate the previous strategies in a comprehensive visual behavior: *The system observes the visual surroundings. If an unexpected residual disparity is detected, the system changes its strategy and foveates the target to analyse until another target is to be observed.* Only one object is taken into account at a given instant.

This behavior can, considering a strategy related to the human oculomotor behavior [1.66], be formalized as follows. We consider two visual fields, in either one or two cameras, one a peripheral field, the other being a foveal field. The first is used to detect potential targets, the second one to track one target and analyse it:

Process for the peripheral visual field $\mathcal{P}$: visual observation and image stabilization.

- Detect unexpected residual disparity and update the average depth and size of each target, choose the target (if any) which residual disparity is maximal.
- If no target is detected:

- Rotate smoothly the camera to minimize the rotational disparity between two frames (*stabilization*) [$\mathcal{R}_1$].
- If the camera is too eccentric, reset the camera position in one step, using a pure rotation, and cancel the visual computations during this motion (*reseting saccade*) [$\mathcal{R}_3$].
- Else, if one target is detected:
  - Rotate smoothly the camera to maintain the target around the principal point of the camera (*tracking*) [$\mathcal{R}_2$].

Process for the foveal visual field $\mathcal{F}$: foveation and 3D object observation.

- If a target is detected:
  - Modify in one step the eye angular position. If the target is in the peripheral part of the visual field, relocate it in the central part of the visual field *(capture and correcting saccade)* [$\mathcal{R}_3$].
  - Keep the observed area near the center of the retina, using smooth eye movements and maintain the size of the object to about half of the size of the retina controlling zoom *(smooth-pursuit)* [$\mathcal{R}_1$]/[$\mathcal{R}_5$].
  - Smoothly move the neck so that the neck orientation corresponds to the average gaze position, over an adaptive time window, decreased to zero when the camera eccentricity reaches its maximum *(eye–neck cooperation)* [$\mathcal{R}_4$].
  - Perform 3D object observation.
- If no target detected:
  - Zoom back to a minimal focal.
  - Perform the same displacements as the other sensor.

The implementation of this behavior depends on the mechanical hardware and will be given after the presentation of an instant ciation of these general ideas, in the next chapter.

2. 3D Active Vision on a Robotic Head

The next step is now to verify whether these general ideas can be easily implemented, considering an effective active visual system. This is the goal of this second chapter in which we describe a mechanical system which has been designed to facilitate the fixation and tracking of 3D objects.

We intend to build a vision system that will allow dynamic 3D perception of objects of interest. More specifically, we discuss the idea of using 3D visual cues when tracking a visual target, in order to recover some of its 3D characteristics (depth, size, kinematic information). The basic requirements for such a 3D vision module to be embedded on a robotic head are discussed.

The mechanics on the mechanical setup described first and a framework to deal with eye–neck control is proposed. The visual mechanisms able to detect visual targets and effectively realize their 3D observations are described in the second part of this chapter. An important issue is that all these mechanisms are adaptive in the sense that strategies which automatically adapt their parameters are proposed.

2.1 A One-to-One 3D Gaze Controller

The RobotVis robotic head is a device capable of controlling the gaze direction of a binocular stereo system, and of tuning the parameters of one lens [2.1]. A front view of the system is shown in Fig. 2.1, and diagram of the top view in Fig. 2.2.

Basically, the system uses a minimum number of three degrees of freedom (pan, tilt and vergence on one eye) for directing gaze on a 3D point without any redundancy. The mechanical relation between the three axes is known, pan and tilt correspond to two orthogonal intersecting axes, while the vergence axis is aligned with the pan axis. Pan axis is on tilt. Using such a kinematic chain has two consequences: it allows a very simple computation of inverse kinematics and since we have very few joints, it increases positioning precision.

The two visual sensors behave differently and they do not have the same characteristics. One is the "foveal sensor" (dominant eye), it is fixed with respect to the mount and has a controlled zoom, focus, and iris. The other is a "peripheral sensor", capable of vergence with a fixed lens. Inertial sensors

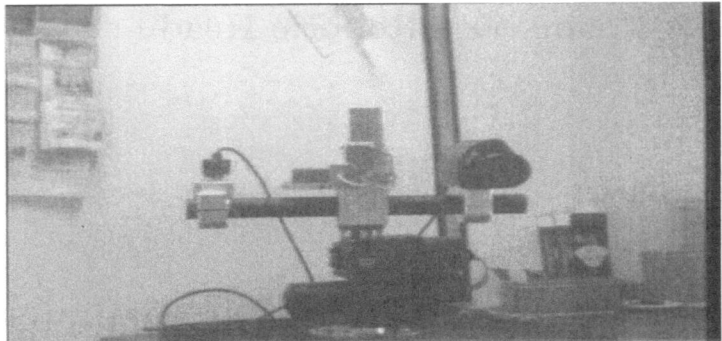

Fig. 2.1. A view of the robotic head

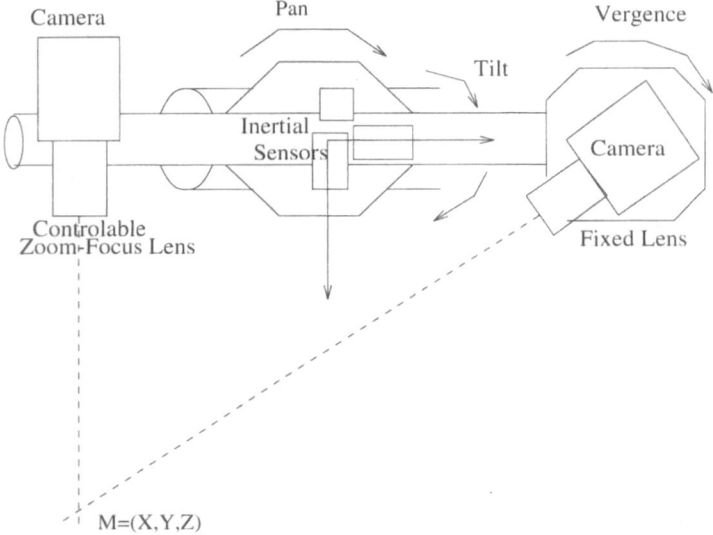

Fig. 2.2. Robotic head mechanical architecture

are also available. This system corresponds to a mechanical instantiation of the proposed framework, an can be considered as the "simplest" robotic head on which 3D vision can be performed.

2.1.1 Technical Data on the Robotic Head

The technology of the system is based on DC-motors and gear boxes with incremental angular sensors controlled by PID loops tuning the angular positions of the system. Local constant acceleration trajectories are automatically generated for any move, all this at 3 kHz, using dedicated VLSI circuits. With such a technology, the overall precision is better than 0.1 deg, and can move up to 60 deg/s. Vergence can reach up to 200 deg/s.

The first camera has a controllable zoom which allows the field of view to span from 45 deg to 3 deg. The second camera has a fixed field of view of about 45 deg. This seems to be the upper limit for the validity of the common linear model of the camera (pin-hole model).

Both cameras have high resolution 449×579, rectangular pixels of 15.18×8.28 μm, the sensibility is 2.0 lux, noise signal ratio is of about 56 dB, parasite reflections are avoided, and the signal gain is automatically controlled within a range of 12 dB. The integration time is adjustable from 20 ms down to 1 ms.

Cameras are synchronized so that the two video frames do not have interferences. We work with a 512×256 picture from a European CCIR-standard 50 Hz interlaced video signal. Image data is provided at 25 Hz every 40 ms, but the two fields of 512×256 pixels are processed separately in each frame, thus providing a 50 Hz data rate. This is done to overcome the blurring induced by interlaced video images. Input data becomes available after 20 ms delay. This configuration is compatible with actual vision machines such as the European (Depth from Motion Analysis) DMA machine [2.2] partially used in this system.

On this robotic system, two types of inertial information are used: (1) Low cost linear accelerometers have a precision of about 0.5 % with a measurement scale between $\pm 2 \times 9.81$ m/s^2, while the resolution is of 0.1 % in a 0–50 Hz range. After calibration [2.3] they provide an estimate of the specific forces. (2) Angular rate sensors which provide information about angular velocity with a resolution better than 0.04 deg/s for the low cost units. Their overall precision is similar to the linear accelerometers. After calibration [2.3] they provide an estimate of the angular velocity.

Let us now discuss system calibration. Although it is established that modifying lens zoom and focus will affect all intrinsic calibration parameters, this effect is rather stable and can be approximately pre-calibrated. We thus can perform several off-line calibrations for different configurations using a standard method [2.4, 5] and obtain the relations between: (1) the intrinsic calibration parameters, zoom ϕ_Z and focus ϕ_F, the depth Z_0 corresponding to the minimum blur for each focus position ϕ_F, (2) extrinsic parameters: eye rotation R_e, neck rotation R_n and vectorial location of the eye with respect to the neck d_{ne}. This completely defines the calibration of the mount.

2.1.2 Computing the Inverse Kinematic for 3D Fixation

Let us derive the inverse kinematic for this robotic head. Several mechanical constraints are verified on this system: pan and tilt axis intersect and are orthogonal, the vergence axis is parallel to the tilt axis, pan is on tilt, the camera optical centers and the three rotation axes are coplanar. Moreover, for the camera with vergence, because the optic is fixed it is possible to have the optical center contained in the vergence rotation axis, and thus this degree of freedom is a pure rotation. However, the relative orientation is not easily controllable, but is known from calibration. The transformation from the mount to the camera, here limited to a rotation, will be noted H_{left} and H_{right}.[1] In practice, for the camera with a fixed lens, $H_{\text{left}} \simeq I$. On the other hand, the camera without vergence but having a variable lens has not only its relative orientation but also its relative position not easily controllable although known from calibration. In practice, this transformation is of the form

$$H_{\text{right}} \simeq \begin{pmatrix} I & (0,0,\Delta_{\text{f}})^{\text{T}} \\ (0,0,0) & 1 \end{pmatrix}$$

since the main effect of a zoom is to translate the optical center along the optical axis z to a quantity equal to the variation of the focal length Δ_{f}, which is a realistic model for high quality cameras [2.6] while it is only an approximation for low quality lenses [2.7].

These two transformations are known from calibration, and their values are expected to be small. They are constant over time, except the translation induces by the zoom. Moreover, it is always possible to compensate for the rotational part of H_{left} and H_{right} by applying a reprojection, that is a linear transform of the homogeneous coordinates of the picture [2.8], and to calibrate these parameters using accurate calibration techniques [2.9]. Finally a fine mechanical adjustment can compensate for the translational components of these static transformations. We thus can consider, from now on, that they can be canceled.

Using the notations of Fig. 2.3, we can compute the kinematic chain in homogeneous coordinates for both cameras:[2]

$$\text{Right camera:} \quad H_{\text{r}} = \begin{pmatrix} R(y,\mathcal{P}) \cdot R(z,\mathcal{T}) & \begin{matrix} -D \\ H \\ \Delta_{\text{f}} \end{matrix} \\ (0,0,0) & 1 \end{pmatrix},$$

---

[1] We use the very common notation:
$$\begin{vmatrix} X' \\ Y' \\ Z' \\ 1 \end{vmatrix} = \underbrace{\begin{pmatrix} R & t \\ (0,0,0) & 1 \end{pmatrix}}_{H} \cdot \begin{vmatrix} X \\ Y \\ Z \\ 1 \end{vmatrix}$$
to relate the coordinates of a point between two frame of references which differ by a rotation R and a translation t.

[2] $R(u, \theta)$ is the matrix of a 3D rotation around an axis u ($\|u\| = 1$) with angle θ. We have explicitly, using the notation $\tilde{X} \cdot u = X \wedge u$:
$$R(u, \theta) = Id_{3x3} + \sin(\theta)\tilde{u} + [1 - \cos(\theta)]\tilde{u}^2 \, .$$

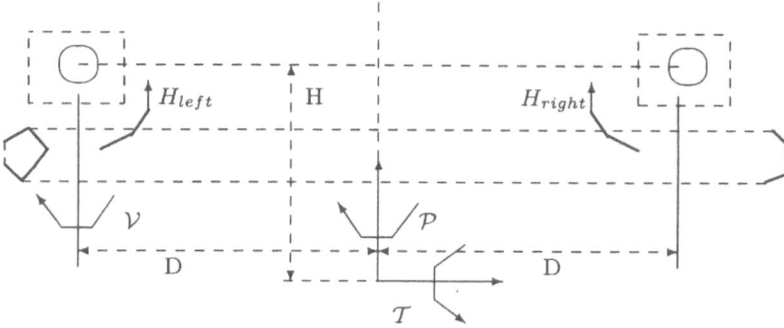

Fig. 2.3. Notations for the head kinematic. $\mathcal{P}$ is the pan angle, $\mathcal{T}$ is the tilt angle, $\mathcal{V}$ is the vergence angle, D the eccentricity of the cameras with respect to the rotation center, H the height of the cameras with respect to the rotation center

$$\text{Left camera:} \quad H_1 = \left(\begin{array}{cc|c} R(y, \mathcal{V} + \mathcal{P}) \cdot R(x, \mathcal{T}) & R(y, \mathcal{V}) \cdot & \begin{array}{c} D \\ H \\ 0 \end{array} \\ (0, 0, 0) & 1 & \end{array} \right).$$

Let us now compute the inverse kinematic. It is generally not possible to gaze at a given point $M = (X, Y, Z, 1)$ in space, with the left camera having vergence but not the right camera. This is because the two optical axes are required to be in the same plane. This however is the case in our configuration. First, M should belong to the plane of the two optical axes. The projection should also belong to the two optical axes.

Solving these previous equations, to fixate a point with world coordinates (X, Y, Z), our robot must be in absolute position angles $\mathcal{P}, \mathcal{V}, \mathcal{T}$:

For bounded angles we can explicitly solve these equations:[3]

$$\begin{cases} \mathcal{P} & = \arctan\left(-\dfrac{X}{Z}\right) \\[2mm] \mathcal{V} & = \arcsin\left(\dfrac{B \cos(\mathcal{P})}{Z}\right) \\[2mm] \mathcal{T} & = 2\arctan\left(\dfrac{B \cos(\mathcal{V})}{-Y \sin(\mathcal{V}) - H \sin(\mathcal{V})} \right. \\[2mm] & \quad \left. \pm \dfrac{\sqrt{B^2 \cos(\mathcal{V})^2 + Y^2 \sin(\mathcal{V})^2 - H^2 \sin(\mathcal{V})^2}}{-Y \sin(\mathcal{V}) - H \sin(\mathcal{V})}\right) \end{cases} \tag{2.1}$$

where the admissible domain is given by

$$\left(\sqrt{Y^2 + Z^2} > H\right)\left(\sqrt{(\cos(\mathcal{T}) Y - \sin(\mathcal{T}) Z)^2 + X^2} > D\right).$$

---

[3] Please note that the equation: $a \cdot \cos(x) + b \cdot \sin(x) + c = 0$ with $\theta = \arctan(a/b)$ can be rewritten as $\sin(x + \theta) = -c/\sqrt{a^2 + b^2}$ thus having at most one solution in $]-\Pi/2, \Pi/2[$ which is $x = -\arcsin(c/\sqrt{a^2 + b^2}) - \arctan(a/b)$. We have a unique solution if and only if a and b are not null together and if $|c|$ is bounded by $1/\sqrt{a^2 + b^2}$, else there is no solution.

These equations provide a direct positioning of the system towards a 3D target, and the computation is fairly easy to implement. It is thus possible to compute in one step the positioning of the mount. This means that we can decouple the problem of the robotic control of this head (feedback, local trajectory generation, etc.) and its visual control. The visual control directly defines angular join positions.

Please note that the equations are independent of Δ_f, i.e. the shift related to the zoom/focus of the lens. There is thus no need to calibrate the corresponding parameter at this stage. This nice property is due to the chosen mechanical architecture.

The low-level control is thus decoupled on each axis and is implemented in hardware, using pre-programmed position or velocity PID loops.

2.2 Active Observation of a 3D Visual Target

We would now like to design the following behavior, on this artificial oculo-motor system: *The system is observing the visual surroundings, and the visual sensor is stabilized in terms of retinal slip. If an unexpected motion is detected the system changes its strategy and tracks the target until it disappears.* Only one moving object is taken into account here.

In order to implement such an algorithm we need two modules: (1) A *controller* which, given a retinal error in position, will drive the visual sensor angular location in order to perform the visuo-motor task.[4] In tracking mode, it corresponds to the target retinal location, in the stabilization mode it corresponds to the displacement of the scene between two frames. In both cases, the input to this module is a retinal error in position, the output being a command to the mount. (2) A *motion-detector* which, given the sequence of pictures, detects the regions corresponding to moving areas and decides what is the target if any, and what is the average retinal slip between two frames. This module input is the image sequence and its output a retinal error to the controller.

Using this very simple architecture we can attempt to implement the previous behavior.

In this section we will only present the first module (controller), a paradigm for motion detection will be discussed in the next section. Several other researchers also attacked this problem of motion detector and have implementations of such a module [2.10–13].

---

[4] Even at the lower levels of the controller, velocity feedback loops are used to control eye/neck motions.

2.2.1 A 1D Model of the Eye–Neck Tracking

In order to analyse this problem, and considering the geometry of our system, we consider a mono-dimensional eye–neck system as shown in Fig. 2.4.

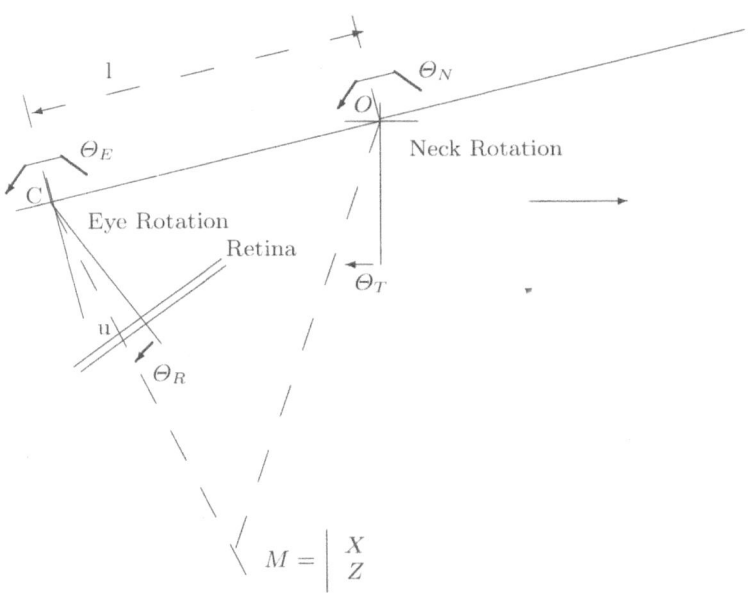

Fig. 2.4. Geometrical configuration of a 1D eye–neck system

Let us take a fixed frame of reference (x, z) with its origin O at the center of rotation of the neck. For zero angles, the eye center of rotation is at $C = (l, 0)$.

Let $M = (X, Z)$ be a pointlike target in motion in front of the system.

The variables are the following:

θ_N is the angular position of the neck;

θ_E is the angular position of the eye;

$\theta_T = \arctan(X/Z)$ is the angular position of the target in space;

θ_R is the angular location of the projection of the target onto the retina.

It is possible to relate those quantities since the kinematic chain of this system is:

$$\begin{vmatrix} X' \\ Z' \end{vmatrix} = \begin{pmatrix} \cos(\theta_E) & -\sin(\theta_E) \\ \sin(\theta_E) & \cos(\theta_E) \end{pmatrix}$$
$$\cdot \left(\begin{pmatrix} \cos(\theta_N) & -\sin(\theta_N) \\ \sin(\theta_N) & \cos(\theta_N) \end{pmatrix} \cdot M + \begin{vmatrix} l \\ 0 \end{vmatrix} \right).$$

Using the usual perspective model to relate the retinal coordinate $u = \tan(\theta_R)$ to the target location $u = fX'/Z' + u_0$ we have

$$\theta_R = \arctan\left(f\frac{X'}{Z'} + u_0\right),$$

where f is the focal length and u_0 the location of the optical center. We can relate θ_R to the coordinates (X', Z') of the target in a frame of reference attached to the retina.

We obtain:

$$
\begin{aligned}
\theta_R &= \arctan\left(f\frac{\cos(\theta_E + \theta_N)\tan(\theta_T) - \sin(\theta_E + \theta_N) + \cos(\theta_E)l/Z}{\sin(\theta_E + \theta_N)\tan(\theta_T) + \cos(\theta_E + \theta_N) + \sin(\theta_E)l/Z}\right.\\
&\qquad\left. + u_0\right)\\
&= \mathcal{F}_{\{f, l, u_0\}}(\theta_T, \theta_E, \theta_N)
\end{aligned}
\tag{2.2}
$$

and this very simple model shows the characteristics of the problem:

- The relation depends upon the unknown depth Z of the target.
- The relation depends upon the not necessarily known characteristics of the camera, difficult to obtain especially when zooming, the focal length f being changed.
- The relation is highly nonlinear, especially when not considering a 1D model, but the complete kinematic chain of an artificial head with more than one degree of freedom and a complete model of the camera.

We thus need to introduce some simplifications in the previous model. The obtained equation will be used for the simulations.

2.2.2 A Linearized Adaptive 1D Model of the Eye–Neck Tracking

Instead of trying to use this equation we would like to use a *linear adaptive technique*, that is to linearize the previous equation and *learn and adapt* the right coefficients of this equation, during tracking .

More precisely, we propose to use the following linearized relation:

$$\theta_R = k\,\theta_T + \Gamma_E\theta_E + \Gamma_N\theta_N$$

in which k, θ_T, Γ_E and Γ_N are unknown, while θ_E and θ_N are measured. Since k and θ_T cannot be recovered independently but only their product we set $k = 1$. The two "gains" Γ_E and Γ_N are to be tuned but they are expected to be locally constant, while we can compute their initial value from (2.2) as

$$
\Gamma_E = \frac{\left.\dfrac{\partial\mathcal{F}(\theta_T, \theta_E, \theta_N)}{\partial\theta_E}\right|_{\theta_T = 0, \theta_N = 0}}{\left.\dfrac{\partial\mathcal{F}(\theta_T, \theta_E, \theta_N)}{\partial\theta_T}\right|_{\theta_E = 0, \theta_N = 0}} = 1
$$

and

$$\Gamma_N = \frac{\left.\dfrac{\partial \mathcal{F}(\theta_T, \theta_E, \theta_N)}{\partial \theta_N}\right|\theta_T = 0, \theta_E = 0}{\left.\dfrac{\partial \mathcal{F}(\theta_T, \theta_E, \theta_N)}{\partial \theta_T}\right|\theta_E = 0, \theta_N = 0} = 1 \ .$$

We consider the motion of the target as a constant acceleration motion, in other words we are going to use a local second order approximation of the target trajectory:

$$\theta_T(t + \Delta_T) = \theta_T(t) + \Delta_T \dot{\theta}_T(t) + \tfrac{1}{2}\Delta_T^2 \ddot{\theta}_T(t)$$

which can be simplified by normalizing the time and taking $\Delta_T = 1$. The choice of a second order model is based on a number of reasons: (1) the command to the actuators is usually related to a torque, thus proportional to acceleration;[5] (2) usual object motions (falling objects, vehicles in motion, etc.) are well modeled taking acceleration into account; (3) since we are tracking bi-dimensional trajectories with curvature, there is a normal acceleration even if the linear velocity is constant. We thus need to introduce acceleration in the model. Well, the human oculomotor system itself, in smooth pursuit, acts as a velocity feedback with a first order predictor using acceleration [2.14].

Finally we assume that, due the mechanical precision[6] of the mount, we can consider θ_E and θ_N as true values.

In fact, the problem to relate the true values of θ_E and θ_N to the measured retinal error θ_R is not trivial. The picture acquisition takes 20 ms and the eye and neck positions at the beginning and at the end of the picture acquisition are not the same. More precisely the location (u, v) of the target on the retina is measured at a time which is a linear function of these coordinates, thus variable in time. In order to overcome this problem, we have to either interpolate these measures or to measure the eye and neck positions, just when the target is located. We implemented the second method, which is more precise and robust, and measure the angular positions θ_E and θ_N in the motion detection module, during the analysis of the picture.

2.2.3 Statistical Filtering of the Linearized Model

Following the previous discussion we have an estimation problem that can be stated as followed: *measuring the target retinal position, we want to estimate the target trajectory, and the different gains of the system.*

Using a Kalman-Filter-based approach [2.15], we can defined our state vector as:

---

[5] The current in a DC motor is proportional to the electric torque.

[6] A typical precision is of 0.01 deg, with a reproducibility better than 0.1 deg.

$$x(t) = \begin{vmatrix} \theta_T(t) \\ \dot{\theta}_T(t) \\ \ddot{\theta}_T(t) \\ \Gamma_E \\ \Gamma_N \end{vmatrix} \ .$$

This vectorial state will be known with an uncertainty represented by an additive white Gaussian noise of covariance S. We want to estimate the state and the covariance of the state.

Our assumptions about its evolution are, taking the previous discussion into account, the following:

$$x(t + \Delta_T) = \begin{vmatrix} \theta_T(t) + \dot{\theta}_T(t) + \frac{1}{2}\ddot{\theta}_T(t) \\ \dot{\theta}_T(t) + \ddot{\theta}_T(t) \\ \ddot{\theta}_T(t) \\ \Gamma_E \\ \Gamma_N \end{vmatrix} = \begin{pmatrix} 1 & 1 & \frac{1}{2} & 0 & 0 \\ 0 & 1 & 1 & 0 & 0 \\ 0 & 0 & 1 & 0 & 0 \\ 0 & 0 & 0 & 1 & 0 \\ 0 & 0 & 0 & 0 & 1 \end{pmatrix} \cdot x(t) \ ;$$

written as $x(t + \Delta_T) = A \cdot x(t)$ the gains Γ_E and Γ_N being considered as locally constant.

The measure θ_R is related to our state by the linear relation

$$\begin{aligned} m &= \theta_R(t) = \theta_T(t) + \Gamma_E\theta_E(t) + \Gamma_N\theta_N(t) \\ &= \begin{pmatrix} 1 & 0 & 0 & \theta_E(t) & \theta_N(t) \end{pmatrix} \cdot x(t) \end{aligned}$$

written as $m = M \cdot x(t)$ and is known with a certain uncertainty related to an additive white Gaussian noise of variance V_R.

The original equations for a Kalman-Filter [2.16–18] related to our problem are

$$\begin{aligned} \hat{x}(t + \Delta_T) &= A \cdot x(t) \ , \\ \hat{S}(t + \Delta_T) &= A \cdot S(t) \cdot A^T + V \end{aligned}$$

and

$$\begin{aligned} x(t + \Delta_T) &= \hat{x}(t + \Delta_T) + K \cdot [m(t) - M \cdot \hat{x}(t + \Delta_T)] \ , \\ S(t + \Delta_T) &= \hat{S}(t + \Delta_T) - K \cdot M \cdot \hat{S}(t + \Delta_T) \end{aligned}$$

with

$$K = \hat{S}(t + \Delta_T) \cdot M^T \cdot \left(V_R + M^T \cdot \hat{S}(t + \Delta_T) \cdot M \right)^{-1}$$

where V is the covariance related to the evolution uncertainty.

As one can see, the algorithm is done in two steps: (1) From an estimate $(x(t), S(t))$, we first predict the evolution of the state vector and its covariance as $\left(\hat{x}(t + \Delta_T), \hat{S}(t + \Delta_T) \right)$ and also add to the covariance the covariance related to the evolution uncertainty. (2) Given a new measure $m(t)$ we correct the predicted value $\left(\hat{x}(t + \Delta_T), \hat{S}(t + \Delta_T) \right)$ using this new information.

However, in our situation we would like to modify this original framework in three ways: (1) Since the measure is mono-dimensional, the variance V_R is not a matrix but a simple scalar. Therefore, these equations can be simplified. (2) Since we want to use this estimate to track a target and cancel its displacement we do not want to estimate the actual value but to *predict* the next value to occur in order to drive the eye and neck accordingly. We thus are going to use a schema with correction followed by prediction, instead of prediction and correction, which is entirely valid since the equations commute. (3) We do not have any information about the quantitative validity of our evolution model since the equations to relate our linear equation to the true model are not well-defined. We thus do not know how to tune the covariance related to the evolution uncertainty, the matrix V.

To overcome this last difficulty, we are going to use a common trick in the field [2.15, 17]. We are going to update the predicted covariance as

$$\hat{S}(t + \Delta_T) = \alpha \left[A \cdot S(t) \cdot A^T \right]$$

with $\alpha \geq 1$, instead of $\hat{S}(t + \Delta_T) = A \cdot S(t) \cdot A^T + V$. This mechanism is not a hack but has a theoretical interpretation: In the Kalman-Filter, all the knowledge about the Markov process related to the statistical model which is behind this mechanism is contained in the estimate $(\boldsymbol{x}(t), S(t))$. It means that the filter "memorizes" all the information from the beginning of its history. In such a case the covariance decreases with time, and if $V = 0$ tends to zero. If $V \neq 0$ it tends to a steady-state value corresponding to the point of equilibrium. At this point of equilibrium, the knowledge lost from one sampling to another because the evolution is not perfectly known, is recovered by the information obtained from the new measurement. It is not always optimal to let the filter "memorize" all the information from the beginning of its history, because it will not "readapt" if the parameters of the system are changed. In that case we would like the system to have a finite memory, i.e. information is vanishing when time is running. If this decay is exponential with time we just obtained the previous equation involving α. This is called an adaptive Kalman-Filter or a finite memory filter and it just corresponds to our situation since the gains and the target dynamic characteristics might change with time.

Recomputing the equations with these modifications yield

$$\hat{\boldsymbol{x}}(t + \Delta_T) = A \cdot \hat{\boldsymbol{x}}(t) + \frac{m(t) - M \cdot \hat{\boldsymbol{x}}(t)}{V_R + M \cdot \hat{S}(t + \Delta_T) \cdot M^T} A \cdot \hat{S}(t) \cdot M^T ,$$

$$\hat{S}(t + \Delta_T) = \alpha A \cdot \left[\hat{S}(t) - \hat{S}(t) \cdot M^T \cdot V_R^{-1} \cdot M \cdot \hat{S}(t) \right] \cdot A^T$$

as can be obtained after a few algebra.

This mechanism will be initialized with:

$$\hat{\boldsymbol{x}}(0) = \begin{vmatrix} 0 \\ 0 \\ 0 \\ 1 \\ 1 \end{vmatrix} \quad ,$$

the covariance matrix $\hat{S}(0)$ being a diagonal matrix with huge terms on the diagonal, as usually done.

The algorithm has been automatically generated from a symbolic calculator.

Given a set of retinal angular positions θ_{R}, corresponding to the eye and angular positions $(\theta_{\mathrm{E}}, \theta_{\mathrm{N}})$, it is thus possible to recover and predict the target position θ_{T} and an estimate of the gains Γ_{E} and Γ_{N} in the system. The whole mechanism is tuned by only two parameters: the retinal variance V_{R} and the adaptation parameter α.

2.2.4 Controlling the Neck and Eye Positions

Using the previous estimate it is possible to drive the eye and neck, in order to cancel the retinal "error" θ_{R}. Using a Kalman-Filter in a controlled loop is something rather common especially in active tracking [2.15]. More precisely, the separation principle [2.19] allows us to compute the optimal command of the system as the combination of an optimal estimator and an optimal controller which yield an optimal feedback [2.18]. In other words, the theory ensures the obtainability of a stable mechanism, given the model is valid.

In our case, since we want to cancel θ_{R}, we want to output the next eye and neck angular positions $(\hat{\theta}_{\mathrm{E}}, \hat{\theta}_{\mathrm{N}})$ such that

$$0 = \hat{\theta}_{\mathrm{T}}(t + \Delta_{\mathrm{T}}) + \hat{\Gamma}_{\mathrm{E}} \hat{\theta}_{\mathrm{E}}(t + \Delta_{\mathrm{T}}) + \hat{\Gamma}_{\mathrm{N}} \hat{\theta}_{\mathrm{N}}(t + \Delta_{\mathrm{T}}) \ ;$$

the values $\hat{\theta}_{\mathrm{T}}(t + \Delta_{\mathrm{T}})$, $\hat{\Gamma}_{\mathrm{E}}$ and $\hat{\Gamma}_{\mathrm{N}}$ being estimated from the filter.

In fact this control is redundant and we have to decide how to balance the command between eye and neck. Due to the definition of the tasks, there is a very simple way to define our control, since it is required that *the neck angular position is either fixed or corresponds to the average value of the eye angular position over a certain period of time.* This constraint can be express using a first order recursive low pass filter such as

$$\hat{\theta}_{\mathrm{N}}(t + \Delta_{\mathrm{T}}) = \gamma \, \hat{\theta}_{\mathrm{N}}(t) + (1 - \gamma) \, \hat{\theta}_{\mathrm{E}}(t + \Delta_{\mathrm{T}}) \ ;$$

this filter has the following infinite impulse response:

$$\hat{\theta}_{\mathrm{N}}(t + \Delta_{\mathrm{T}}) = \sum_{i=-\infty}^{0} (1 - \gamma) \gamma^{|i|} \hat{\theta}_{\mathrm{E}} \left[t + \Delta_{\mathrm{T}}(i + 1) \right]$$

and is obviously a normalized low-pass exponential filter; its complex gain is
$G(\omega) = (1 - \gamma) / \left[1 - \gamma \exp(-\mathrm{j}\omega \Delta_{\mathrm{T}}) \right] \ .$

The "6 dB" time window, or half-period, of this filter defined as the time after which the average value of the input signal vanishes to more than 50 %,[7] is thus given by

$$\gamma = \left(\frac{1}{2}\right)^{1/(1+T_2/\Delta_{\rm T})}$$

and it is consequently possible to adjust γ for a given time window.

Can we stop the neck, or the eye? With $\gamma = 0$ the neck is stationary, since its angular position remains constant, with $\gamma = 1$ the neck is performing all the motion. We thus only need one control mechanism for all cases.

Combining the two equations, we get the final algorithm which corresponds to the following nonlinear controller:

$$\hat{\theta}_{\rm N}(t + \Delta_{\rm T}) = \frac{\gamma\,\hat{\theta}_{\rm N}(t) - (1 - \gamma)\,\hat{\theta}_{\rm T}(t + \Delta_{\rm T})}{1 + (1 - \gamma)\,\hat{\Gamma}_{\rm N}/\hat{\Gamma}_{\rm E}} \,,$$

$$\hat{\theta}_{\rm E}(t + \Delta_{\rm T}) = -\frac{\hat{\theta}_{\rm T}(t + \Delta_{\rm T}) + \hat{\Gamma}_{\rm N}\hat{\theta}_{\rm N}(t + \Delta_{\rm T})}{\hat{\Gamma}_{\rm E}} \,.$$

This is indeed a "sub-optimal" control mechanism since we do not derive the equations from a chosen criterion to be minimized, but from a simple heuristic. We will however see that this yields reasonable results.

2.2.5 Automatic Tuning from the System Residual Error

In the previous mechanism, we have to adjust the retinal error variance $V_{\rm R}$, the adaptive factor α, and the balance ratio between eye and neck motion γ.

In some situations we are willing to tune these parameters "manually", mainly in two situations: (1) Cancelling neck motion ($\gamma = 1$). (2) Reseting the filter when a new target is detected or when we consider its internal state must be recomputed or in order to perform fast motions. In such cases, we do not filter the target trajectory but simply *saccade the gaze toward the target*. In every case we have to set $\alpha = \infty$ for one sampling period then $\alpha = 1$. This will increase the covariance and put the filter in its initial mode. An equivalent feature is to reset the covariance matrix to its initial value.

But most of the time we do not want to have the duty to adjust these parameters, and it is important to design an upper-layer for this controller capable of automatically tuning these three values. What can such a supervisor observe? Obviously, the imperfection of the system: its residual error.

Let us, then, analyse the system residual error

$$\epsilon(t + \Delta_{\rm T}) = \hat{\theta}_{\rm T}(t + \Delta_{\rm T}) + \hat{\Gamma}_{\rm E}\theta_{\rm E}(t + \Delta_{\rm T}) + \hat{\Gamma}_{\rm N}\theta_{\rm N}(t + \Delta_{\rm T}) \,.$$

It can have three origins:
(1) The true command has not been executed exactly by the mount. In that

---

[7] It is given by $\frac{1}{2} = \sum_{i=-T_2/\Delta_{\rm T}}^{0}(1 - \gamma)\gamma^{-i}$, as easily calculable.

case, the predicted values given to the mount the last sampling period do not correspond to the true values obtained one sampling period later. This first partial residual error is thus equal to the error, on the retina, due to the difference between the desired command and the effective output:

$$\epsilon_1(t + \Delta_T) = -\hat{\Gamma}_E \left[\theta_E(t + \Delta_T) - \hat{\theta}_E(t) \right] - \hat{\Gamma}_N \left[\theta_N(t + \Delta_T) - \hat{\theta}_N(t) \right] .$$

(2) The target dynamic does not correspond to the internal (2nd order) model of the filter, meaning that the estimation is biased. The related partial residual error is thus equal to the difference between measured and predicted values:

$$\epsilon_2(t + \Delta_T) = \left[\hat{\Gamma}_E \theta_E(t + \Delta_T) + \hat{\Gamma}_N \theta_N(t + \Delta_T) \right]$$
$$- \left[\theta_T(t) + \Delta_T \dot{\theta}_T(t) + \tfrac{1}{2}\Delta_T \ddot{\theta}_T(t) \right] .$$

(3) The retinal position of the target is estimated with a certain error (measurement noise of variance V_R). Since there are no other sources of error, and since these errors accumulate, we have

$$\epsilon_3(t + \Delta_T) = \epsilon(t + \Delta_T) - \epsilon_1(t + \Delta_T) - \epsilon_2(t + \Delta_T)$$
$$= \theta_T(t + \Delta_T) + \Delta_T \dot{\theta}_T(t + \Delta_T) + \tfrac{1}{2}\Delta_T \ddot{\theta}_T(t + \Delta_T)$$
$$+ \hat{\Gamma}_E \hat{\theta}_E(t) + \hat{\Gamma}_N \hat{\theta}_N(t) .$$

2.2.6 Estimating V_R

The last source of error is only related to the measurement imprecision and its variance can thus be taken as an estimator of the measurement variance:

$$\hat{V}_R = E[\epsilon_3^2] .$$

This variance has to be computed "off-line", in a time-window large enough not to perturbate the real-time process. This restriction given, we have a way to automatically tune one of the parameters of the system, without having to control it. It is very important not to feedback this parameter at each period in the filter because oscillations will occurr: increasing this parameter will immediately decrease the correction done by the filter using new measurements. This will thus increase $\epsilon(t + \Delta_T)$ then $\hat{V}_R$ and so on. However, tuning this parameter over a long time period will not lead to such oscillations since this effect is local and compensated as soon as the state covariance is updated (theoretically 3 iterations for this 2nd order model, since the state is entirely "observed" from three independent measurements).

2.2.7 Estimating α

If the dynamics of the trajectory do not correspond to the target motion we will have an increase in the second error ϵ_2 and we then better increase α to make the system adapting. On the contrary, if the mount is not following the target properly, we have to keep considering a constant acceleration, since it is related to the motor command and does not allow the system to adapt to a new acceleration until the motors have reached the correct speed, we thus better decrease α. In a saccadic mode α has to be high as already pointed out. After several experimental comparisons we found that a realistic criteria of evaluation of α is

$$\alpha = \log \left(1 + \frac{E[\epsilon_2^2]}{\max(E[\epsilon_1^2], E[\epsilon_3^2])} \right) .$$

Using this relation, α will increase if and only if the error due to the target motion prediction becomes higher than the two other sources of error. It will not increase if either the true motion is not executed by the mount, or the measurement error is high.

2.2.8 Estimating γ

The neck motion has to compensate for eye motion especially if the eye is not at its center position. We thus can modify the factor γ from 0 if the eye is centered to 1 if the eye is eccentric. A reasonable angular interval in which the eye displacements can be performed is $\Theta_E^{\max} \simeq 20$ to 40 deg. Moreover, if the eye moves with an average velocity $E[\dot\theta_E]$, the drift between the eye and the neck will be related to the time window of the filter used and the average eccentricity might be approximated by $\delta \simeq 2E[\dot\theta_E] T_2$ as discussed previously. It is thus possible to adjust γ in order to have this eccentricity below $\Theta_E^{\max}$.

Requiring that the eccentricity remains below $\Theta_E^{\max}$ its admissible value is thus given by the relation

$$|E[\theta_E]| + 2|E[\dot\theta_E]|T_2 < \Theta_{\text{MAX}}$$

and we have

$$2 \left[\frac{\log(\gamma)}{\log(2)} + 1 \right] |E[\dot\theta_E]|\Delta_T < \Theta_{\text{MAX}} - |E[\theta_E]|$$

which allows to derive γ.

2.2.9 Simulation of the Combined Behavior

We have evaluated our algorithm using synthetical noisy data. We have implemented a simulation of the nonlinear mono-dimensional model of the eye–neck system shown in Fig. 2.4 and corresponding to (2.2). We then have verified the performances of the algorithm with respect to this model. Typical results are shown in Fig. 2.5 and Fig. 2.6. This software allows to simulate the dynamical and adaptive behavior of the algorithm.

In fact, most experimental aspects of this system can be observed on the given plot: At the beginning of the estimation, the system internal estimate

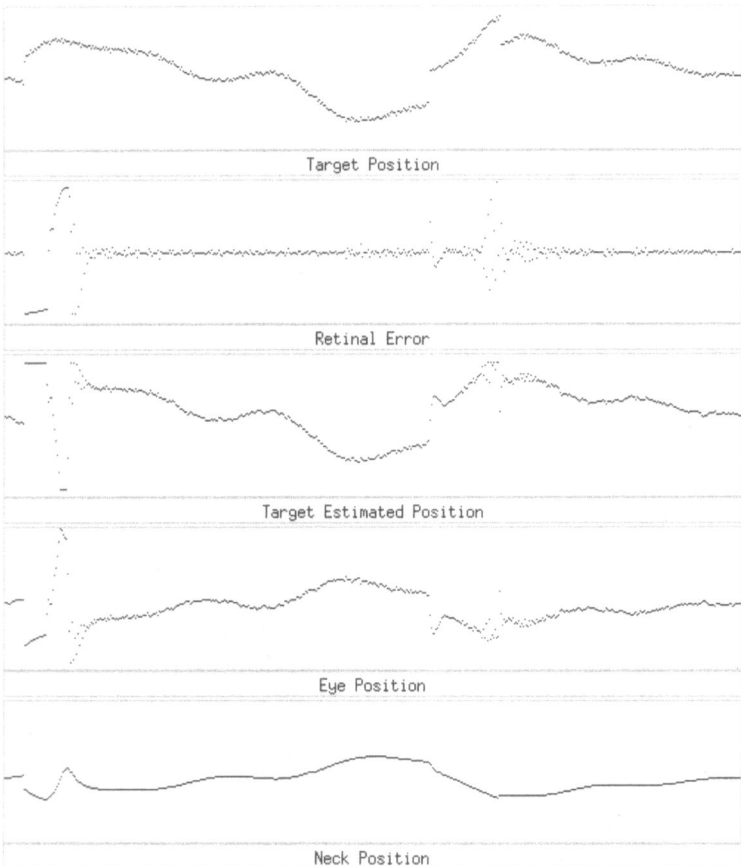

Target Position

Retinal Error

Target Estimated Position

Eye Position

Neck Position

Fig. 2.5. A typical tracking sequence using synthetical noisy data

is unstable because of the important retinal error (the linear model is not valid). However, despite this effect, the eye and neck motions are stable, due to the filtering process. Moreover, in practice such effects are expected to be limited by the mechanical inertia. When a saccade occurs in the steady-state mode, the system performs a quick motion reducing the retinal error, as expected. It is clear that such a perturbation introduces a short overshoot in the eye motion, while the increase of the variances yields a temporarly unstable phase with respect to the noise. The compensatory motion appears to be well balanced between the eye and neck degrees of freedom, and our simple strategy seems to correspond to what is expected to occur in practice, for such a redundant system. Due to the use of the acceleration in our model, the system is capable to track trajectories with complex dynamics as shown

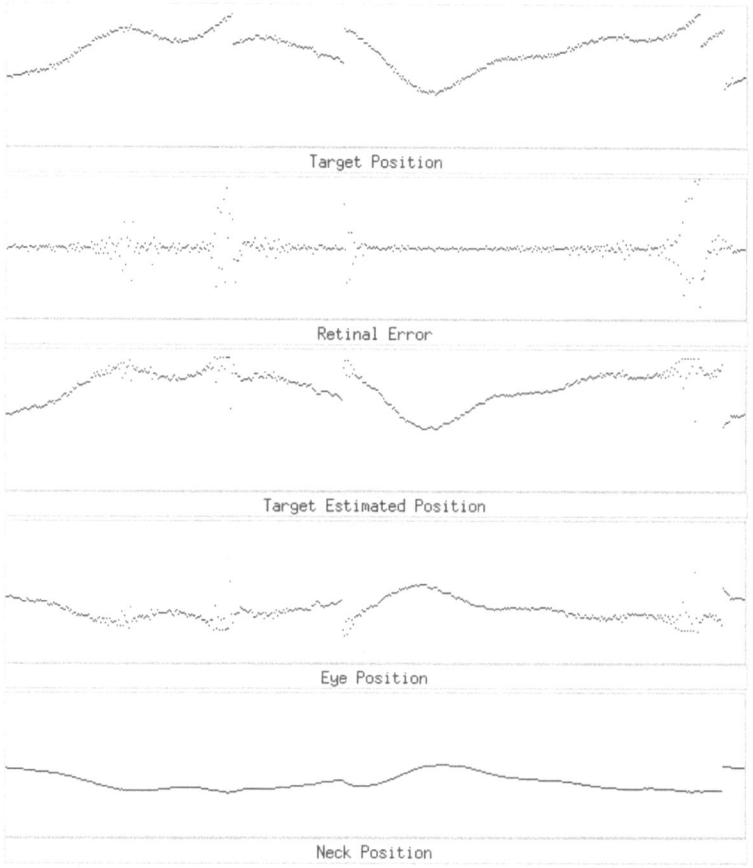

Fig. 2.6. Another typical tracking sequence using synthetical noisy data

in the picture. This is an important improvement with respect to standard $\alpha - \beta$ trackers which only predict the target position and velocity.

Conclusion. We conseqently can tune all the parameters of the controller either automatically, or through specific commands (saccade request, neck stop, etc.) in order to perform specific oculomotor tasks.

These values are to be computed using average estimates ($E[\ldots]$) at a time-scale which is large enough not to interfere with the filter. A reasonable value is about 50 times the sampling period.

2.3 Detection of Visual Targets for 3D Visual Perception

As explained in the previous chapter, when considering 3D vision as a goal, active visual strategies can be divided into two classes:

(1) Strategies to detect visual targets to be submitted to 3D visual perception, i.e. **where to look next?**
(2) Strategies to maintain and improve the 3D perception of the visual targets, i.e. **how to track a visual target?**

Let us analyse these two problems, the first in this section, the other in the next section.

Considering attention focusing we can easily give a list of objects in a scene which are to be preferred when observed by an active visual system: (1) Moving objects might trigger potential alarms ("prey", "predator", moving obstacles to avoid). (2) Nearby objects will be the first to interact with the system as potential obstacles. Their proximity should ease their 3D observation. (3) Objects with a high density of edges correspond to informative parts of the visual field and might be worth a closer look.

These three categories might appear as very natural, indeed they are, but in addition, they correspond to a very precise 3D motion property: *considering that the 3D rotational disparity has been canceled between two consecutive frames (i.e. the motion disparity between is only induced by a translational motion), the residual disparity is only due to (1) objects in motion, (2) objects with a non-negligible depth[8], (3) objects with complex texture or shapes, likely to be detected with some errors, and whose related disparity is not correctly detected.*

In other words, these structures correspond to points with a non negligible residual disparity, the global rotational disparity of the scene being cancelled. The mechanism developed, for instance by Murray et al. [2.20], is based implicitly on this principle although the authors do not clarify this point. So this mechanism of focus of attention on an area of interest is very simple to design. Let us describe an algorithm based on this idea.

Using a Stable Image Without Rotational Disparity. Using the odometric cues, for a calibrated system, we cancel the rotational disparity between two frames. In our system, the rotational displacement is measurable using inertial and odometric cues [2.3, 8] and can easily be canceled considering a collineation of the retinal plane [2.8]. The input is two consecutive frames, the rotational displacement between those two frames, and the intrinsic calibration parameters of these two frames. The output consists of is two consecutive frames, the previous frame being stabilized with respect to the present one. This mechanism has been extensively described elsewhere [2.8, 21].

---

[8] It is well known that the retinal motion amplitude for a given target due to a translational motion is proportional to the ratio between the translation velocity and the target depth.

Segmenting the Image in Terms of Regions of Homogeneous Intensity. We then have chosen to detect areas of homogeneous intensity. Using a very rapid algorithm of region segmentation, the "toboggan method" [2.22], we group and extract regions with constant intensity. The input of this module is one stabilized frame, a parameter to choose the smoothing factor and a parameter to adjust the threshold under which intensity variation is taken as negligible. The output is a list of regions with homogeneous intensity, each region being characterized by its average retinal (1) horizontal and (2) vertical locations, (3) its size, and (4) the average intensity. These four numbers will be used in the sequel.

Matching Regions Between Two Frames. Now, in order to compute the 3D characteristics of these "regions" some other parameters must be estimated. First, since we want to compute depth from focus we must estimate the relative amount of blur for a given region. This will be estimated from (5) the average of the intensity gradient magnitude as in [2.23], the summation being taken over the N pixels of the region. Then, since we want to estimate depth from motion we also need to find (6) the correspondent of the region in the previous frame. We thus have implemented a region tracker, but because of the stabilization mechanism between two consecutive frames, we have limited our implementation to finding the closest region, with a similar average intensity. This restrained but fast implementation was sufficient in practice for not very textured and relatively simple images [2.24].

Auto-Tuning of the Early-Vision Parameters. Any parameter defined in a system must be either tuned automatically or kept fixed, since no manual adjustment can occur. In this module we have two parameters: (1) a smoothing factor, since each image is smoothed before the region segmentation is performed; (2) a gradient threshold, since the boundary between two regions corresponds to a variation in the intensity, i.e. a non-negligible gradient magnitude.

Let us discuss how to automatically compute these two parameters: we have experimented that, on one hand, when the smoothing factor is very small the number of regions increases because some noisy parts of the disparity map are not filtered and appear as regions. On the other hand, when the smoothing factor is rather high the number of region increases also because the disparity map tends to be flattened and some bounds between regions are canceled. We thus expect these two complementary effects to yield an optimal value of the smoothing factor. This value simply corresponds to the smoothing factor for which the number of regions is minimum, which can easily be obtained. The gradient threshold can be considered as the limit between two statistical distributions, one containing the points with a negligible contrast and is expected to be zero up to a certain level of noise, the second contains points with a residual disparity. Using over-classical methods [2.25, 26] we have implemented an automatic detection of this limit. These mechanisms are run once at the beginning of a session, during a boot-strapping phase. They are not run continuously because they can perturbate the other estimators since they modify the number of regions and their shapes.

Detecting Regions of Unexpected Disparity. Finally, we detect whether the residual retinal disparity of the center of each region is less than a given threshold. As before, this threshold is computed automatically by considering the limit between two statistical distributions, one corresponding to regions with negligible disparity, the other corresponding to regions of interest as in [2.26].

The corresponding visual module can be summarized as followed:

(a) Cancel rotational disparity between two frames.
(b) Detect regions of homogeneous intensity.
(c) Match each region with the "closest" region in the previous frame.
(d) Adapt the internal parameters, if in the bootstrapping phase.
(e) Detect region(s) of non-negligible residual retinal disparity.

Visual Module [$\mathcal{V}_1$]: Where To Look Next

If more than one region has been detected, the system must decide which one is to be observed first. However, in our experimentations, since we do not have to perform any specific task we have simply chosen the area with the highest residual disparity to be observed first.

2.4 Computing the 3D Parameters of a Target

In a reactive system, depth can be obtained from several cues: stereo disparity, motion disparity, zoom disparity, blur variations and focus. These different cues do not have the same precision [2.27]. Moreover, we expect the two first ones to provide only an information about the "average depth" of the object. In order to deal with this situation we propose to analyse the 3D structure in two steps: we first consider the object as a flat fronto-parallel shallow of constant depth and estimate its depth and motion, and then refine this 3D estimation using stereo disparity only, since recovering the 3D structure of an object in motion is often a difficult problem. On the contrary, stereo mechanisms can be used during motion [2.28], since for a moving object, the 3D structure can be estimated at each instant without knowing its motion.

Since the observed object is, here, assumed to be of constant depth, the projection of the center of gravity of the points of the 3D object corresponds to the center of gravity of the projections of the points of the 3D object as easily verified (Fig. 2.7). Therefore, we can compute the depth for the center of gravity of the object only. This is done as follows.

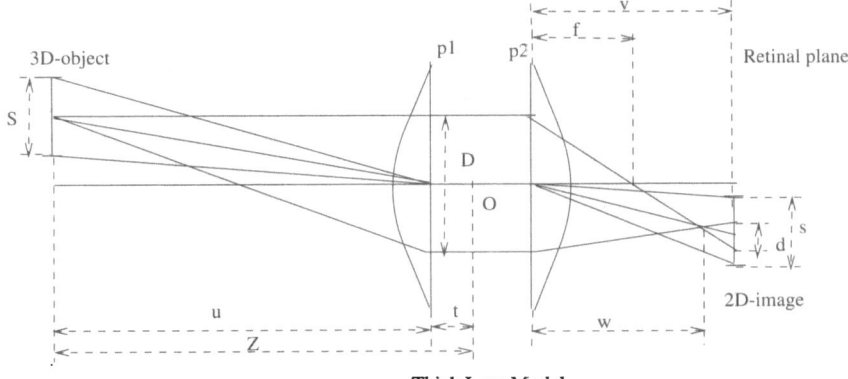

Fig. 2.7. The camera geometry and the object geometry; $p1$ and $p2$ are the principal planes, D is the aperture of the lens, f the focal length; the image formation is done at a distance w of the principal plane, and the retina is at a distance v of the retina. The 3D object size is S and its distance to the lens Z, it is the sum of the distance u from the object to the principal plane $p1$ and the distance t from $p1$ to the origin. The 2D object size is s and the blur circle diameter d

2.4.1 A Unique Framework to Integrate Depth from Vergence, Motion and Zoom

Let us consider a Euclidian frame of reference attached to the mount and a 3D point $M = (X, Y, Z)^{\mathrm{T}}$ in this frame of reference. The retinal projection $m_i = (u_i, v_i)^{\mathrm{T}}$ of this point onto a camera $\mathcal{C}_i$ can always be written [2.29]:

$$\lambda_i \begin{pmatrix} u_i \\ v_i \\ 1 \end{pmatrix} = P(\boldsymbol{\theta}) \cdot M + \boldsymbol{p}(\boldsymbol{\theta}) \tag{2.3}$$

where $P(\boldsymbol{\theta})$ is a 3×3 matrix function of the mount and lens parameters (angular positioning of the mount, zoom, focus) $\boldsymbol{\theta}$, $\boldsymbol{p}(\boldsymbol{\theta})$ a 3 dimensional vector also function of the same parameters. The P-matrix and $\boldsymbol{p}$-vector are known from calibration and contain the intrinsic and extrinsic parameters of the mount and camera. In particular, the disparity induced by zooming is integrated in these equations.

As soon as the object has been located in an image, the three homogeneous equations (2.3) yield, λ_i being eliminated, two linear equations as a function of the 3D location of the center of gravity of the object. Having correspondences between two frames in a binocular system provides at each instant 4 linear equations to recover this 3D location.

2.4.2 Using Second-Order Focus Variations to Compute Depth

Moreover, if the system has a calibrated auto-focus module, each time the system is focused for a given retinal location we have an estimation of the distance from the point to the optical center of the camera Z_i, because there is a direct relation between the target depth and the distance at which we are focused (with the notations of Fig. 2.7, we have from the lens formula $1/f = 1/u + 1/w$: $Z_i = u + t$ as a function of w).

But if the system is not focused, from the geometry of the lens, we can calculate the diameter of the blur circle [2.23] as

$$d = D\,v\,\left(\frac{1}{f} - \frac{1}{u} - \frac{1}{v}\right).$$

This value can be related to the intensity distribution but the obtained results, according to almost every researcher are very qualitative and need some additional calibration procedures. Moreover, as stated by Anderson [2.30] there is a twofold ambiguity, since the blur exists whenever the retina is behind or in front of the image plane. We thus need at least two measures to distinguish these two configurations. It has also been established that the relations between depth and blurr from out of focus are deeply dependent upon calibration parameters of the lens, which are quite difficult to obtain.

Following another approach, we have interpolated the point of optimal focus using the following simple **local** model, which is not obtained from the physical model of the lens:

$$L_i(\boldsymbol{\theta}) = \log(\sigma_I^2) = A + B\,[W_i(\boldsymbol{\theta}) - Z_i]^2$$

The quantity $W_i(\boldsymbol{\theta})$ is the distance corresponding to an object on focus, for a mount configuration $\boldsymbol{\theta}$. This model is justified by the fact the intensity variance is approximately related to the focus by a Gaussian function [2.31]. The quantities A and B are unknown. Now considering three values of the intensity variances, at time 0, 1 and 2, we can eliminate A and B and obtain

$$Z_i = \frac{1}{2}\;\frac{\begin{array}{l}W_i(\boldsymbol{\theta}_0)^2\,[L_i(\boldsymbol{\theta}_2) - L_i(\boldsymbol{\theta}_1)] \\ + W_i(\boldsymbol{\theta}_1)^2\,[L_i(\boldsymbol{\theta}_0) - L_i(\boldsymbol{\theta}_2)] \\ + W_i(\boldsymbol{\theta}_2)^2\,[L_i(\boldsymbol{\theta}_1) - L_i(\boldsymbol{\theta}_0)]\end{array}}{\begin{array}{l}W_i(\boldsymbol{\theta}_0)\,[L_i(\boldsymbol{\theta}_2) - L_i(\boldsymbol{\theta}_1)] \\ + W_i(\boldsymbol{\theta}_1)\,[L_i(\boldsymbol{\theta}_0) - L_i(\boldsymbol{\theta}_2)] \\ + W_i(\boldsymbol{\theta}_2)\,[L_i(\boldsymbol{\theta}_1) - L_i(\boldsymbol{\theta}_0)]\end{array}}$$

which allows to estimate Z_i even if not focused. More precisely if Z_i is not constant we obtain an average value of Z_i at time 1. This model is, of course, only valid when close to the focus.

The variance attached to this equation can be related to the peak of the focus, i.e. the parameter A, also called "depth of focus" and is easily obtained from [2.27], (25). Das and Ahuja have computed the variance of this quantity as a function of the lens parameters. But if we make the following realistic hypotheses $Z \gg t$, $Z \gg f$ and $D \gg d$ which corresponds to a remote object and an image almost on focus, we simply have:

$$V_{Z_i} \simeq k\,\frac{f^2}{D^2}.$$

This means that the precision of the focus is proportional to the focal length and inversely proportional to the aperture. The factor k has been calibrated on the lens.

2.4.3 Multi-Model Concurrency in 3D Tracking

We thus have, at each instant, 6 measurement equations: 2 coming from focus in both lens, 2×2 coming from the target location in both cameras. Obviously, if one or more of these measures are undefined we just can avoid their integration by considering their (co)variance as infinite (the inverse of the (co)variance is zero). In other words, each of these 6 measurement equations are weighted by the inverse of a variance. So if only one camera is used or if no information about depth from focus is available, the estimator can still be run with the corresponding weights equal to zero.

These measurement equations have been used in a set of linear Kalman filters. Four filters are running in parallel. Each filter estimates the center of gravity of 3D object $M = (X, Y, Z)$. The first filter assumes M is stationary, i.e. its velocity is null. The second filter assumes M is moving with a constant velocity, and the third filter assumes M is moving with a constant acceleration. An additional filter assumes that the previous location of M is unknown and does not rely on prior information. Assuming a Gaussian additive noise, for each filter a probability of error is computed from the residual error, related to a Ξ^2 distribution. The state of the system (M location, velocity and acceleration) is updated considering the best filter output. This well known mechanism of multi-Kalman filtering [2.15] allows to detect whether the target is either moving at constant acceleration, or moving at constant velocity, or whether it is stationary.

Moreover, because of the last filter we can detect model ruptures. If the internal state is longer reliable, the last filter – which does not rely on previous information – will have a better estimation than the three first filters. On the contrary, the three first filters explicitly use the previous state since they integrate some models for the evolution of the parameters. Then, for a target either moving either at constant velocity or acceleration or stationary, one of the first three filters should provide a better fit.

Finally, this adaptive mechanism allows to always use a model with a minimal number of parameters since, for instance, the target acceleration is not estimated if negligible with respect to the system noise.

2.4.4 Considering Further 3D Information

Considering that each pixel has a variance $V = \sigma^2 I$ – i.e. has a constant and isotropic uncertainty –, the variance V_g for the center of gravity of a region of N pixels is obvious to compute: $V_g = I \sigma^2 / N$.

The average 3D size of the object is obtained immediately from the 2D size of the object since: $S = s Z/v$, as can be seen in Fig. 2.7.

Moreover, we can use this average depth to initiate reconstruction algorithms. Let us discuss this point now. Each time a visual target is detected its size and depth is computed and the target can be reconstructed as a fronto-parallel object. Moreover, not only the target structure but also the target 3D motion is analysed. Since this is the result of a Kalman filtering, the quality of the results improves as the number of equations increases. Therefore the system must maintain the observation of these visual targets. Putting all these targets together, allows to build *a coarse dynamic depth and kinematic map of the surroundings*. Several 3D vision modules can refine this estimation, but the actual process allows to select, sort and detect the data input. One experimentation of this kind has been reported in [2.24].

Let us describe the related implementation:

(a.1) Considering the gradient magnitude variations in three views we generate an estimation of the distance from the target to the optical center of each camera projected on the optical axis, with its variance, eventually for both cameras.
(a.2) Considering the center of gravity of the object image in eventually both cameras generate four equations about the target 3D location.
(b) Using the measurement equations update the four Kalman filters and compute the residual error. Update the state of the system considering the filter with a minimal probability of error.
(c) Compute the average size of the target.

Visual Module [V₂]: Average Depth and Size Recovery

Although rather rudimentary, this 3D information is sufficient to drive the mount and maintain the observation of a 3D object. This is demonstrated in the next section.

2.5 Experimental Results

2.5.1 Head Intrinsic Calibration

In the first part of this experimental work, we have elaborated a simple model for the variations of the lens intrinsic parameters when the zoom and focus parameters are modified.

Our analysis demonstrates that a linear model is sufficient to describe these modifications. Experimentally, this approximation is valid, with the precision being better than 1.5 pixels in almost every case, and 2–3 pixels at most.

We have obtained these numbers by a set of repetitive static calibrations using a well established interactive modern method of calibration reported in [2.9].

An example of such a curve is given in Fig. 2.8. The complete set of results is available in [2.7].

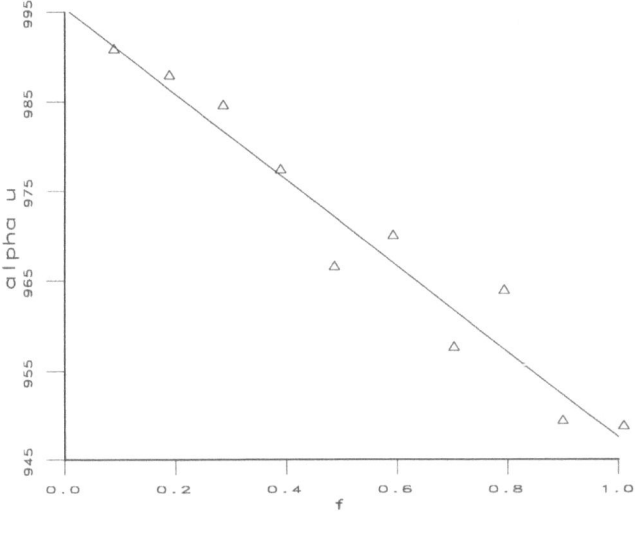

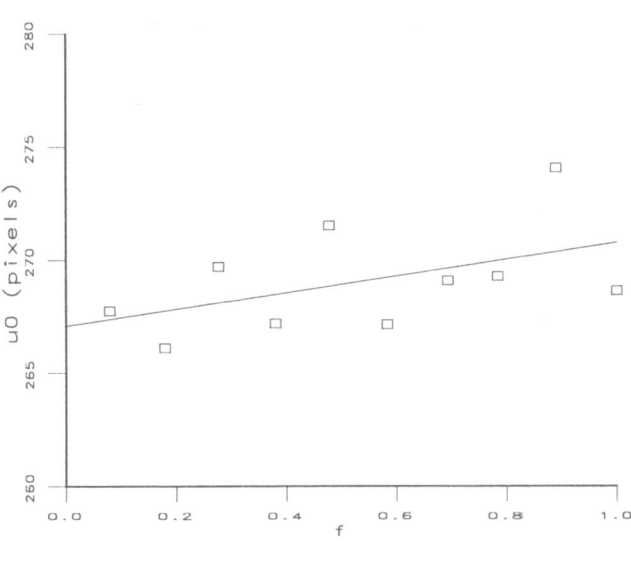

Fig. 2.8. Stability of the horizontal intrinsic calibration parameters during a change of focus. The first trace is the horizontal scale factor and the second trace is the horizontal location of the principal point. The precision is of about 2–3 pixels and very similar results have been obtained for vertical parameters and during a zoom

2.5.2 Looking at a 3D Point

We then have verified the quality of the head calibration by gazing at several locations in space. We have compared the location at which the head was really gazing with the expected location. We have manually measured the 3D locations, therefore the precision of this experiment is no more than 0.5 cm.

We also have noted the retinal error induced by the uncertainty in the calibration, a sample is given in the following table:

Measured (X, Y, Z) [m]			Expected (X_0, Y_0, Z_0) [m]			Retinal Error (ϵ_u, ϵ_v) [pixel]
(0.000,	0.000,	1.062)	(0.000,	0.000,	1.000)	(0, 1)
(0.112,	0.005,	1.013)	(0.100,	0.000,	1.000)	(4, 2)
(−0.004,	0.098,	1.013)	(0.000,	0.100,	1.000)	(4, 3)
(−0.980,	0.120,	1.998)	(−0.100,	0.100,	2.000)	(2, 3)
(−0.960,	0.110,	2.898)	(−0.100,	0.100,	3.000)	(2, 1)

Although rather approximative, this small experiment yields two conclusions: (1) the precision of the head calibration is quite reliable, the precision being of about 1–2 cm, except for one point; (2) such uncertainty corresponds to retinal errors close to the precision of the lens calibration.

If we have to say it in two words, this system has a precision of "2 cm" and "4 pixels".

2.5.3 Where to Look Next Experiment

The next module to check was the focus of attention module. We have analysed the behavior of this part of the system, when an object of unexpected residual disparity was detected. Such an example is shown in Fig. 2.9.

We have analysed the performance of this module by measuring the retinal motion thresholds corresponding to the detection of an unexpected moving object, as shown here:

Object apparent size [pixel]	279	110	52	23
Motion Threshold [pixel/frame]	4	6	5	4

We also have analysed the retinal motion thresholds corresponding to the detection of a close stationary object, as shown here:

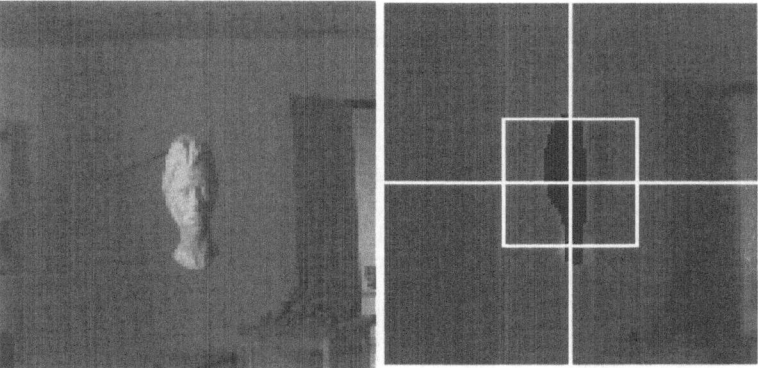

Fig. 2.9. An example of detection of a moving object, while the head was in motion. The *left image* shows the picture, and the *right image* the detected region. This is an area of 279 pixels, with an average intensity variation of 21 over the 255 intensity range, and an apparent motion of 2 pixels

Object apparent size [pixel]	221	113	72	27
Disparity Threshold [pixel/frame]	3	5	4	4

These two results are quite important because they demonstrate two things: (1) the thresholds are close to the actual precision of the system and (2) they do not depend on the object size, which is important for applications.

2.5.4 Reconstruction of a Coarse 3D Map

We then would like to report on the vision modules dedicated to the recovery of a coarse 3D map. This module has been built from already existing algorithms and is not the main part of this work but has also been reported elsewhere [2.24]. As a consequence, we do not need to discuss in great detail the experimental results but we simply provide one example, and give the main properties.

As shown in Fig. 2.10, we have established experimentally the automatic adjustment of the smoothing factor indeed yields a realistic value of smoothing and is close to what an expert would have chosen. The region clustering is not very efficient, because quite a lot of very small regions are detected (Fig. 2.10), but a simple fixed threshold of the region minimal size, allows to eliminate this problem. This is visible in the left part of Fig. 2.11.

The precision of the overall system is good enough to recover a coarse depth map as shown in Fig. 2.11, and to detect residual disparity as analysed in the previous section.

Fig. 2.10. Optimal smoothing of a real moving scene observed during a motion. The objects are on a table with wheels which has been moved during the picture acquisition. The region segmentation is shown on the *right image*, one tip for each region

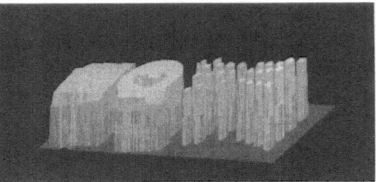

Fig. 2.11. The computation of motion for the best matched regions and the corresponding depth map, for the scene shown in the *previous figure*

2.5.5 Tracking of a 3D Target: Simulation Experiments

In order to verify the validity of our design we have built a simulator to analyse the behavior of the 3D estimator. This simulated system not only predict the 3D location and motion of a punctual target but also detect the class of motion (no motion, constant velocity, accelerations) and model ruptures. This mechanism is comparable to what has been issued in [2.32] on an arm–eye system.

Considering the simulation reported in Fig. 2.12, for instance, it is visible that when the dynamics of the target are changed the system automatically resets and tracks the target with a small phase lag. This was not obvious because the level of noise is quite important (5 pixels) and thus the system has to perform a rather huge filtering.

Moreover we have observed another advantage of this implementation. This is related to stability. Considering an internal model including target acceleration (a second-order filter), it is clear that some instabilities can occur, especially when a prediction of the target is used in a closed-loop mode. This also occurs here, but instantaneously, a more stable model (a first-order filter) is chosen by the algorithm, because it minimizes the measurement error, whereas the unstable filter does not. Therefore, an automatic switch is done to recover a stable feed-back. This is an important property.

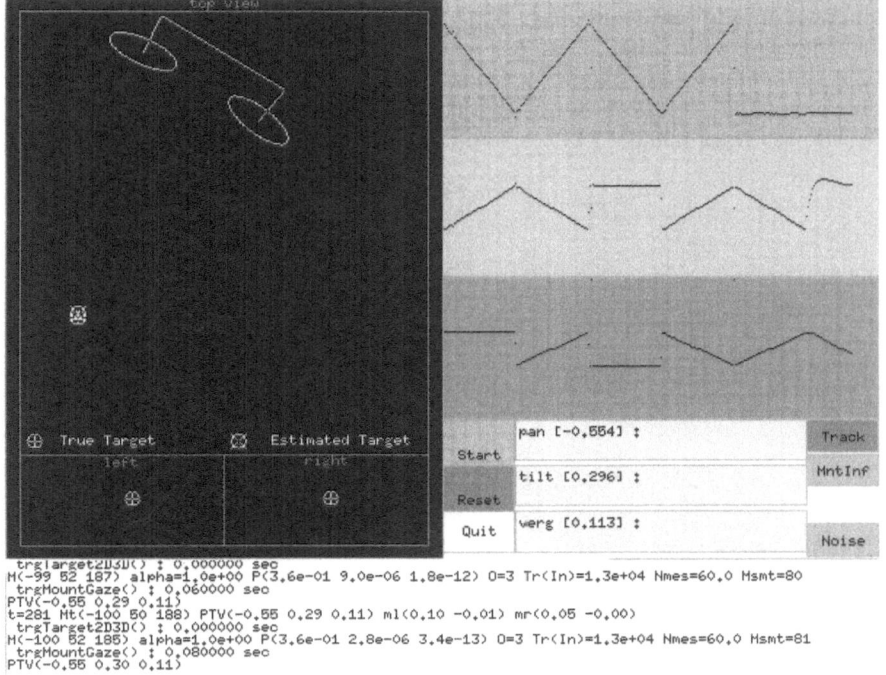

```
trgTarget2D3D()  :  0.000000 sec
H(-99 52 187) alpha=1.0e+00 P(3.6e-01 9.0e-06 1.8e-12) O=3 Tr(In)=1.3e+04 Nmes=60.0 Msmt=80
 trgMountGaze()  :  0.060000 sec
PTV(-0.55 0.29 0.11)
t=281 Ht(-100 50 188) PTV(-0.55 0.29 0.11) ml(0.10 -0.01) mr(0.05 -0.00)
 trgTarget2D3D()  :  0.000000 sec
H(-100 52 185) alpha=1.0e+00 P(3.6e-01 2.8e-06 3.4e-13) O=3 Tr(In)=1.3e+04 Nmes=60.0 Msmt=81
 trgMountGaze()  :  0.080000 sec
PTV(-0.55 0.30 0.11)
```

Fig. 2.12. Simulation of the 3D tracking of a noisy target with abrupt variations in the dynamics. The *top view* represents the binocular mount during the tracking of the target represented as a point. Retinal projections are shown *below* in *left* and *right* cameras. The estimation of the 3D target location is shown *on the right* for X, Y and Z coordinates (*from top to bottom*). Each trace corresponds to 10 s and $5 \times 5 \times 10$ m considering real-time tracking (25 Hz)

2.5.6 Tracking of a 3D Target: Real Object Experiments

We have finally conducted an experiment using the INRIA robotic head and have tracked several targets (objects on a table with wheels, humans, manufactured objects, etc.) at a frequency higher than 1 Hz. Such a sampling rate is, of course, a drastic limitation of the dynamics of the target but our goal was only to realize an experiment, not an industrial system. Due to the nature of the image processing, it was always possible to track the moving object even for retinal disparities close to half the retinal size of the target. We also have verified that the qualitative behavior of the 3D estimator corresponds to what has been obtained during simulations.

The main positive feature is the following: the system is able to (a) predict the target motion and (b) re-center in one step, so that we have observed that the tracking is very robust in the sense that it does not "lose" the target, even if we have significant delays. It also provides a quite efficient tracking as shown in Fig. 2.13 and Fig. 2.14. It is very difficult to compare these performances with other work in the field, because except for one scientist [2.33] the experimental results are limited to a description of the qualitative behavior of the algorithm. On the contrary, even if this is for a limited range of result we have tried to collect some quantitative results.

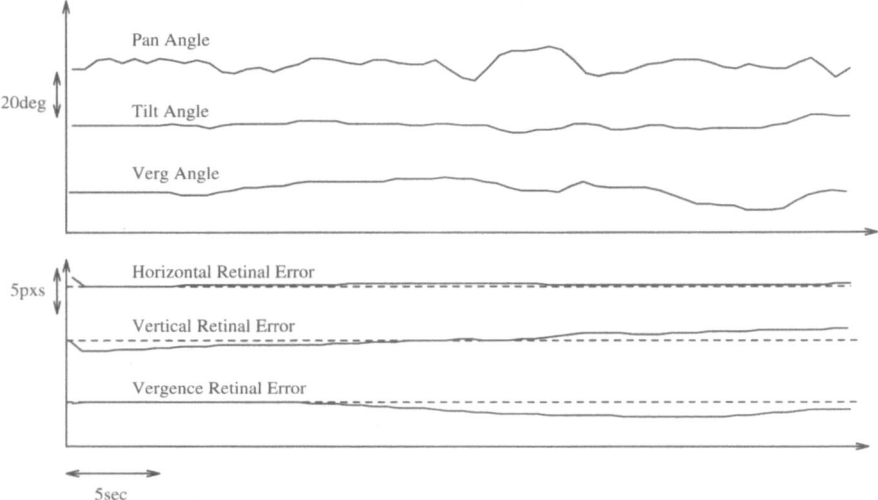

Fig. 2.13. Results of a real 3D tracking on the INRIA head. The angular head positions and the corresponding retinal errors are shown. The average error is of 2–3 pixels and the corresponding image sequence is given in the *next figure*

2.5.7 Conclusion

Let us briefly summarize the results obtained in this chapter.

- We can control the different degrees of freedom of a mount (zoom, focus, vergence, gaze direction) using 3D visual cues, in a single step. The same control in 2D would have to be iterative.
- We can provide the relationship between the mount angular position and the retinal displacement in pixels, and keep the observed object in the foveal part of the cameras, even for slow systems. This can only be done using a 3D calibration model (mount + visual sensor), and computing object depth.
- We detect regions of interest considering the retinal disparity related to the translational component of the motion. Such detection is very similar to principles developed in [2.34] but the use of an active visual system allows to simplify the original procedure.

Fig. 2.14. Results of a real 3D tracking on the INRIA head. One image every four frames is shown, for one camera only. Please note that the system has performed a zoom and has refocused

- We can track moving or stationary tokens in an image sequence by considering a realistic model of their rigid motion. We also determine the nature of the 3D motion of the observed object.
- The average depth and size of the observed objects are computed and a coarse 3D map is reconstructed.

Although this was not obvious at first, the additional complexity of 3D reactive sensing behaviors not only preserves but also further enhances the capabilities of an artificial visual system. It provides a higher level of knowledge representation for task oriented visual systems.

The success of this experimentation is deeply related to the existence of a mechanic, "tuned to the task". This means that we might have to reconfigure a robotic head for each visual application.[9] Moreover, we must make profit of the fact that the high mechanical precision of the system allows to know the calibration of the camera (extrinsic and intrinsic) in any configurations.

---

[9] The INRIA system has the capability of being easily reconfigurable.

3. Auto-Calibration of a Robotic Head

Visual sensor calibration represents the problem of determining the parameters of the transformation between the 3D information of the imaged object in space and the 2D observed image. Such a relationship is mandatory for 3D vision. More precisely, we have to know the location (translation) and attitude (rotation) of the visual sensor with respect to the rest of the robotic system (extrinsic parameters), and the different parameters of the lens, such as focal length, magnitude factors, optical center retinal location (intrinsic parameters).

We now propose a new method of auto-calibration for a visual sensor mounted on a robotic head. The method is based on the tracking of stationary targets, while the robotic system performs a specific controlled displacement, namely a fixed axis rotation. In this chapter we derive the equations related to this paradigm and demonstrate that we can calibrate intrinsic and extrinsic parameters using this method.

3.1 Introduction

The problem of calibrating a number of cameras is becoming extremely important in the field of active vision to make use of artificial robotic heads [3.1, 2].

Even though this problem has been often considered in the past for static visual systems [3.3–5], very little is known about the calibration of active visual systems, in which the extrinsic parameters and the intrinsic parameters parameters of the visual sensor are modified dynamically. A review is given in [3.6]. For instance, when tuning the zoom and focus ratios of a lens these parameters are modified and must be recomputed.

In addition, due to the lack of easily usable calibration procedures, this has lead many workers in the field to base their calibration techniques on the observation of a specific object (calibration grid) which sizes and shape are known in advance. Such a calibration must be done at the beginning of a working session. In the case of an active visual system, calibration of the visual sensor for any possible configuration must be done, which is not realistic since such a mount has 5 to 13 degrees of freedom.

However, some researchers have already attempted to define procedures of calibration in which the equations used to identify the parameters do not come from the knowledge of a given object, but from hypotheses made about the visual surroundings. Such approaches are either based on general projective invariants [3.7, 8], or limited to the problem of calibration of the epipolar geometry of a stereo rig [3.9–11]. In fact, the word "auto-calibration" or self-calibration for a visual sensor is applicable when the method allows the system to calibrate (1) automatically, (2) without any semantic knowledge of the observed objects, (3) without a complete knowledge of the self-motion.

Considering all these methods, the basic constraint is the following: the observed objects are assumed to be stationary, thus forming a rigid configuration. The first question is thus how to detect tokens corresponding to the same rigid object, without the knowledge of their 3D parameters ... without calibration! Perhaps surprisingly, this problem has been solved already in the past considering image motion information [3.12, 13] or image tokens [3.14]. Moreover, although the 2D models used to segment moving objects are not entirely rigorous, they are often more robust than methods involving 3D parameters [3.15], especially for small motion disparities [3.16]. In the case of several moving objects present in the field of view, the segmentation mechanism can easily decide which rigid motion corresponds to the stationary tokens [3.12, 14]. It is then reasonable to base our study on the assumption that the visual input is a set of stationary tokens.

Finally, token correspondences from one frame to another, are required by such mechanisms. This is obtained using token-tracker techniques [3.17, 18]. In this chapter we are going to consider points as tokens.

Specific and possibly more efficient mechanisms can be developed in the caseof a robotic head. Such mechanisms are specific because on a robotic head, we partially know the motion of the visual sensor. More precisely, all robotic heads perform controlled rotations around a few fixed axes, and the relative angle of this rotation is known with high precision[1].

Using this capability, one might expect some simplifications in the problem of calibration. Moreover, there is a need to design a specific method in this case because general methods fail, since this kind of motion is singular for a general method of auto-calibration as shown by Luong [3.19]. In order to overcome this problem, we must introduce an additional information, for instance about the angle of the rotation. We thus will make two assumptions about the active motion of the sensor: (1) the rotation is performed around a fixed axis, and (2) the relative angle of the rotation is known.

Can a system be calibrated using these two assumptions? The answer is yes, and the related equations are developed in this chapter.

---

[1] Typical precision is of the order of 0.01 deg, with a reproducibility better than 0.1 deg for an angular (rotatory) join.

3.2 Reviewing the Problem of Visual Sensor Calibration

We use *the standard pinhole model* for a camera, assuming the camera performs a perfect perspective transform with center O (the camera optical center) at a distance f_0 (the focal length) of the retinal plane.

It must be noted that the pinhole model can still be used for a zoom lens if the object-to-image distance is not considered constant [3.20]. This is the case here, since we will adapt the camera metric for different object locations.

All coordinates are related to an affine frame of reference $\mathcal{R} = (O, \boldsymbol{x}, \boldsymbol{y}, \boldsymbol{z})$ *attached to the retina*, $\boldsymbol{z}$ being aligned with the optical axis, $\boldsymbol{x}$ and $\boldsymbol{y}$ being aligned with horizontal and vertical axis in the image, respectively. The retinal plane is thus perpendicular to the optical axis Oz.

We represent a 3D point $M = \boldsymbol{OM} = (X, Y, Z)^{\mathrm{T}}$ using Euclidian coordinates. Points on the retina, with horizontal and vertical pixel coordinates (u, v), will be represented as homogeneous 3D vectors: $s\,m = s\,\boldsymbol{Om} = (s\,u, s\,v, s)$, corresponding to lines of a given direction passing through the optical center (2D projective space), as shown in Fig. 3.1. This retinal frame corresponds to a rotation angle $\Theta = 0$.

Algebraic developments will thus involve 3D vectors in both cases.[2]

In this study, *we cannot assume that the system is calibrated.* On the contrary, we consider the following well established model for a camera, using an A-matrix, which represents the intrinsic parameters:

$$
\begin{aligned}
Z\,m \;&=\; \begin{vmatrix} Z\,u \\ Z\,v \\ Z \end{vmatrix} \\[6pt]
&=\; \left(\underbrace{\begin{pmatrix} \alpha_u & \gamma & u_0 \\ 0 & \alpha_v & v_0 \\ 0 & 0 & 1 \end{pmatrix}}_{A} \;\; \begin{matrix} 0 \\ 0 \\ 0 \end{matrix} \right) \cdot \begin{vmatrix} X \\ Y \\ Z \\ \mathcal{U} \in \{0,1\} \end{vmatrix} = A \cdot M \,. \quad (3.1)
\end{aligned}
$$

This corresponds, if $\gamma = 0$, to the usual equations $u = \alpha_u\,X/Z + u_0$, $v = \alpha_v\,Y/Z + v_0$. A review is given in [3.6].

For points not at infinity, we have $\mathcal{U} = 1$ and Z corresponds to the Euclidian depth in the retinal frame of reference $\mathcal{R}$ attached to the retina and defined previously. Equation (3.1) is still valid for "points at infinity", i.e. vanishing points, with $\mathcal{U} = 0$, but Z no longer corresponds to the depth of the point.

Very briefly:

---

[2] We denote vectors with **bold** letters and matrices with capital letters. $\tilde{\boldsymbol{u}} \cdot \boldsymbol{x} =$
$$
\begin{pmatrix} 0 & -u_z & u_y \\ u_z & 0 & -u_x \\ -u_y & u_x & 0 \end{pmatrix} \cdot \boldsymbol{x} = \boldsymbol{u} \wedge \boldsymbol{x}
$$
corresponds to the cross-product, the dot-product being written as $\boldsymbol{x}^{\mathrm{T}} \cdot \boldsymbol{y}$. The identity matrix is written I.

- The parameter γ takes into account the fact that, if the retinal plane of the CCD is not orthogonal to the optical axis of the lens, a CCD element (a pixel) will not be projected as an orthogonal rectangle, but as a parallelogram ([3.21], Chap. 2). Most often, for commercial cameras, we can assume $\gamma = 0$ [3.6].[3]
- The 2D point $p = (u_0, v_0)$ corresponds to the projection of the optical center onto the retina (principal point).
- The ratio $\rho = \alpha_u/\alpha_v$ depends upon both the geometry of a pixel and the way the digitisation is performed. It is usually different from one, because pixels are rectangular, and the sampling rates of the digitalization board might not be homogeneous. However, this value is almost fixed, and thus has to be calibrated only once [3.22].
- The average value of α_u and α_v is directly related to the focal length of the lens f_0, and varies when zooming [3.22].

In fact, this "linear projective model" is, for large angles of view, valid only in a certain area of the retina and has to be modified for different retinal eccentricity, as discussed in [3.3]. More precisely such a deviation could be of about 3 %, thus about 5 pixels, for controllable lenses, i.e. small but not negligible [3.20]. Its is also known that this model is valid only for a given 3D depth area. We will thus adapt the parameters depending on the object location, which is required when zooming, as previously mentioned.

In the rest of the chapter we are going to use the matrix A to represent the intrinsic parameters of the system. This matrix is invertible since $\alpha_u > 0$ and $\alpha_v > 0$.

When zooming, or modifying the parameters of the lens, both the magnitude factors and the optical center might change and it is very important to be able to recover the intrinsic parameters $(u_0, v_0, \alpha_u, \alpha_v)$.

Let us now see how to derive a method of calibration which is only based on the observation of stationary points of the background, while performing a partially known motion.

3.3 Equations for the Tracking of a Stationary Point

Let us consider a fixed point $M = (X, Y, Z)$ in space and its projection onto the image plane in homogeneous coordinates $m = (u, v, 1)$. During the head displacement, the point $M = (X, Y, Z)$ does not project directly onto the retina, but is subject to this rotation, say of axis Δ and angle Θ. For $\Theta = 0$ the rotated frame of reference corresponds to the retinal frame of reference.

The axis is parameterized by a unitary vector $u = (u_x, u_y, u_z)^{\mathrm{T}}$ aligned with the axis of rotation and a point $C = (C_x, C_y, C_z)$ on this axis. We choose

---

[3] If f_0 is the focal length, θ the angle (close to $\pi/2$) of the pixel parallelogram, we have from [3.21]: $\alpha_u = -f_0\, k_u, \alpha_v = f_0\, k_v/\sin(\theta), \gamma = -\alpha_u \cot(\theta)$, thus if $\gamma = 0$, we have $\theta = \pi/2$ and $f_0 = (\alpha_u + \alpha_v)(k_u + k_v)$.

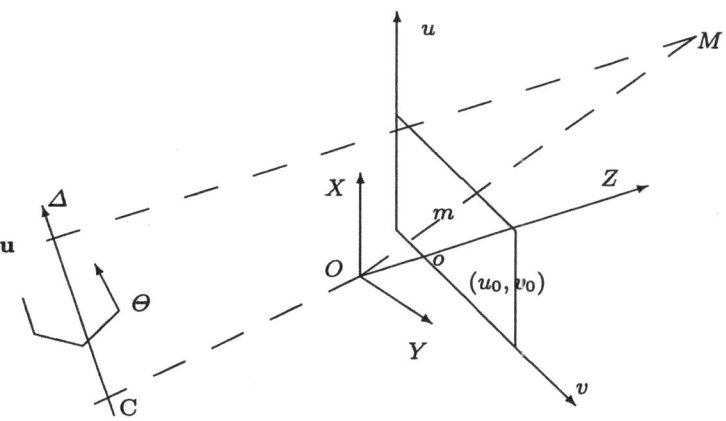

Fig. 3.1. Geometric relationships in the calibration problem

C in order to have $OC \perp u$. Thus, this point is uniquely defined, and $\|OC\|$ is the distance from Δ to the origin. The related equation is:

$$M'C = R(u, \Theta) \cdot MC \Rightarrow M' = R(u, \Theta) \cdot M + \underbrace{\{[I - R(u, \Theta)] \cdot C\}}_{t} \quad (3.2)$$

where $R(u, \Theta)$ denotes the rotation matrix, while the induced translation is t.

The projection of M onto the image plane describes a planar trajectory which can be parameterized by Θ. The coefficients are functions of: (1) the intrinsic calibration parameters $(u_0, v_0, \alpha_u, \alpha_v)$, of (2) the rotation parameters (extrinsic parameters) u and C, and (3) the 3D depth Z of the point.

Let us analyse this relation:

There is an explicit formula for $R(u, \Theta)$, using the notation $\tilde{u} \cdot x = u \wedge x$,[4] which is related to a compact representation of the rotations (Rodriguez formula) [3.23, 24]. We have:[5]

$$R(u, \Theta) = Id_{3 \times 3} + \sin(\Theta)\tilde{u} + [1 - \cos(\Theta)]\tilde{u}^2 . \quad (3.3)$$

Using (3.2) and (3.3) we can write:

$$\begin{aligned} \mathcal{Z}(\Theta)\, m(\Theta) &= A \cdot [R(u, \Theta) \cdot (M - C) + C] \\ &= A \cdot \{M + \sin(\Theta)\tilde{u} \cdot (M - C) \end{aligned}$$

---

[4] For a vector $u = (u_x, u_y, u_x)^{\mathrm{T}}$ we have $\tilde{u} = \begin{pmatrix} 0 & -u_z & u_y \\ u_z & 0 & -u_x \\ -u_y & u_x & 0 \end{pmatrix}$.

[5] This representation is well defined for rotations with $\Theta \in]0, \Pi[$ and $u \in \mathcal{S}^2$ (unit sphere). If $\Theta = 0$ (identity) or $\Theta = \Pi$ (symmetry) this representation is singular, but the formula is still valid. Moreover, since we control Θ this situation can be avoided in practice.

$$+ [1 - \cos(\Theta)] \tilde{\boldsymbol{u}}^2 \cdot (M - C) \}$$
$$= A \cdot \{ \boldsymbol{u} \cdot \boldsymbol{u}^{\mathrm{T}} \cdot M + C + \sin(\Theta) \boldsymbol{u} \wedge (M - C)$$
$$+ \cos(\Theta) \left[(I - \boldsymbol{u} \cdot \boldsymbol{u}^{\mathrm{T}}) \cdot M - C \right] \} .$$

This equation can be factorized as

$$\begin{vmatrix} u(\Theta) \\ v(\Theta) \\ 1 \end{vmatrix} = \frac{Z}{\mathcal{Z}(\Theta)} \left[\boldsymbol{\mathcal{O}} + \boldsymbol{\mathcal{C}} \cos(\Theta) + \boldsymbol{\mathcal{S}} \sin(\Theta) \right] , \tag{3.4}$$

if we write

$$\mathcal{P} = \begin{vmatrix} \boldsymbol{\mathcal{O}} = \begin{pmatrix} O1 \\ O2 \\ O3 \end{pmatrix} = \frac{1}{Z} \left[(\boldsymbol{u}^{\mathrm{T}} \cdot M) A \cdot \boldsymbol{u} + A \cdot C \right] \\ \boldsymbol{\mathcal{S}} = \begin{pmatrix} S1 \\ S2 \\ S3 \end{pmatrix} = \frac{1}{Z} \{ A \cdot [\boldsymbol{u} \wedge (M - C)] \} \\ \boldsymbol{\mathcal{C}} = \begin{pmatrix} C1 \\ C2 \\ C3 \end{pmatrix} = \frac{1}{Z} \left[A \cdot (M - C) - (\boldsymbol{u}^{\mathrm{T}} \cdot M) A \cdot \boldsymbol{u} \right] . \end{vmatrix} \tag{3.5}$$

Considering each image coordinate, the previous set of equations can also be written:

$$u(\mathcal{P}, \Theta) = \frac{C1 \cos(\Theta) + S1 \sin(\Theta) + O1}{C3 \cos(\Theta) + S3 \sin(\Theta) + O3} ,$$

$$v(\mathcal{P}, \Theta) = \frac{C2 \cos(\Theta) + S2 \sin(\Theta) + O2}{C3 \cos(\Theta) + S3 \sin(\Theta) + O3} . \tag{3.6}$$

Considering (3.6), we have the following obvious result.

Proposition 3.1. During a fixed axis rotation, the projection of a stationary point onto the retina is a planar curve – the equation of the planar curve is given in (3.6) – and parameterized by the angle of rotation Θ. Its coefficients are

$$\mathcal{P} = (O1, O2, O3, S1, S2, S3, C1, C2, C3)^{\mathrm{T}} ,$$

uniquely defined up to a scale factor.

As expected, $\mathcal{P}$ is a function of the calibration parameters, which are $\{\boldsymbol{u}, C, (\alpha_u, \alpha_v, u_0, v_0)\}$, and of the point location M, as given by (3.5). Moreover, all the information available about the calibration parameters, considering the projection of a stationary point, is contained in $\mathcal{P}$.

With this formulation, we can perform auto-calibration in two steps:

(1) Identification of the coefficients of the trajectory of the projection of one or several stationary points during a rotation of the mount. This is done in the following section.
(2) Computation, knowing these coefficients, of some information about the calibration parameters. This is done in the section thereafter.

3.4 Recovering the Parameters of the Trajectory

Given a set of angles Θ_i and a set of related retinal points (u_i, v_i), $i = 1, \ldots, n$, we now want to compute the vector $\mathcal{P}$ up to a scale factor as given in (3.6).

In order to choose a criterion for this estimation, we first make use of two simple hypotheses:

– For each data point (u_i, v_i) the measurement errors are independent and have a constant covariance:

$$\Lambda_i = \begin{pmatrix} \sigma^2 & 0 \\ 0 & \sigma^2 \end{pmatrix} .$$

– The quantities $\{\Theta_1, \ldots \Theta_N\}$ are known and their related measurement errors are negligible.

In such a case, a reasonable criterion is:

$$
\begin{aligned}
\xi^2 &= \frac{1}{\sigma^2} \sum_i \left\{ \left[u_i - \frac{C1\cos(\Theta_i) + S1\sin(\Theta_i) + O1}{C3\cos(\Theta_i) + S3\sin(\Theta_i) + O3} \right]^2 \right. \\
&\quad \left. + \left[v_i - \frac{C2\cos(\Theta_i) + S2\sin(\Theta_i) + O2}{C3\cos(\Theta_i) + S3\sin(\Theta_i) + O3} \right]^2 \right\} \\
&= \frac{1}{\sigma^2} \sum_i \left\{ [u_i - u(\mathcal{P}, \Theta)]^2 + [v_i - v(\mathcal{P}, \Theta)]^2 \right\} .
\end{aligned}
\tag{3.7}
$$

This corresponds to a Ξ^2 criterion, also called a pseudo-Mahalanobis distance, and has a statistical interpretation [3.24]. However minimizing this criterion leads to nonlinear equations, and it might not be straightforward to find the solutions.

In order to overcome this difficulty we are going to recover the coefficients of the trajectory in two steps. We first are going to derive analytically an initial estimate for the coefficients, and then refine this first estimate numerically.

3.4.1 An Initial Estimate of the Coefficients

A simple way to obtain an initial estimate of the coefficients is to compute the Taylor expansion of these expressions. Let us only consider u_i, the same derivations hold for v_i. The Taylor expansion for u_i, up to the third order is a huge expression but can, in fact, be worked out if we use the constraint $O3 + C3 = 1$, which is true (see Prop. 3.3). This constraint is useful to simplify the equations, and insures the expression not to be ill-conditioned for $u_i|_{\Theta_i = 0} = (O1 + C1)/(O3 + C3)$. In that case we obtain

$$u_i = (C1 + O1)$$
$$+ (S1 - S3\,C1 - S3\,O1)\,\Theta_i$$
$$+ \left(\frac{O1}{2} - \frac{C1\,O3}{2} - \frac{O1\,O3}{2} - S3\,S1 + S3^2\,C1 + S3^2\,O1\right)\Theta_i^2$$
$$+ \left(\frac{S1}{3} - \frac{S3\,C1}{3} - \frac{5\,S3\,O1}{6} - \frac{S1\,O3}{2}\right.$$
$$\left. + S3\,C1\,O3 + S3\,O1\,O3 + S3^2\,S1 - S3^3\,C1 - S3^3\,O1\right)\Theta_i^3$$
$$+ o(\Theta_i^4)$$
$$= u_{i0} + u_{i1}\,\Theta_i + u_{i2}\,\Theta_i^2 + u_{i3}\,\Theta_i^3 + o(\Theta_i^4)$$

and a similar expansion exists for v_i.

Considering this Taylor expansion up to the third order allows us to compute $\mathcal{O}$, $\mathcal{S}$ and $\mathcal{C}$ from a set of linear equations. Using the same notations for v we obtain, after a small amount of algebra,

$$\begin{cases} u_{i1}\,O3 + 2\,u_{i2}\,S3 &=& 2/3\,u_{i1} - 2\,u_{i3} \\ v_{i1}\,O3 + 2\,v_{i2}\,S3 &=& 2/3\,v_{i1} - 2\,v_{i3} \\ C3 &=& 1 - O3 \\ O1 &=& 2\,u_{i2} + u_{i0}\,O3 + 2\,u_{i1}\,S3 \\ S1 &=& u_{i1} + u_{i0}\,S3 \\ C1 &=& -2\,u_{i2} + u_{i0}\,C3 - 2\,u_{i1}\,S3 \\ O2 &=& 2\,v_{i2} + v_{i0}\,O3 + 2\,v_{i1}\,S3 \\ S2 &=& v_{i1} + v_{i0}\,S3 \\ C2 &=& -2\,v_{i2} + v_{i0}\,C3 - 2\,v_{i1}\,S3 \end{cases} \qquad (3.8)$$

The first three equations allow us to compute the common denominator between u and v from linear equations, the last set of equations to compute the numerators.

These equations are well defined if and only if $u_{i1}\,v_{i2} - u_{i2}\,v_{i1} \neq 0$. A singular estimate corresponds to the case where the curvature of trajectory $\kappa = 2\,(u_{i1}\,v_{i2} - u_{i2}\,v_{i1})/\left(u_{i1}^2 + v_{i1}^2\right)^{3/2}$ is zero, that is, a rectilinear trajectory. We have determined experimentally that this corresponds to the case where the axis of rotation is parallel to the retinal plane while the projection of the tracked trajectory is close to the center of the retina. This situation can thus be avoided in practice.

A first estimation of the trajectory can thus be obtained considering a polynomial model, and using – for instance – least square minimization.

The polynomial models fitted to the u and v data do not need to be of order three, but can depend on the nature of the trajectory.

As soon as one polynomial model for u or v is at least of order two and the other one at least of order one the set of equations (3.8) is well-defined.

3.4.2 Minimizing the Nonlinear Criterion

Using the previous algorithm to get an initial estimate, the components of $\mathcal{P}$ are finally obtained minimizing the criterion given in (3.7).

In the proposed implementation, the minimization is performed using a comprehensive algorithm for finding the minimum of a sum of squares [3.25]. No derivatives are required, which is an advantage (less computations, adaptive estimation of either the Gauss–Newton directions or the Newton directions depending on the numerical stability of the algorithm, etc.) over methods requiring the analytic gradient. The method is designed to ensure that steady progress is made whatever the starting point, and to have a rapid ultimate convergence, i.e. super-linear.

In order to estimate these nine coefficients up to a scale factor, at least four data points are required. This means the point must be tracked in at least four views.

Let us now compute the covariances for the coefficient $\mathcal{P}$.

This is a nonlinear least-square minimization, corresponding to fitting a parametric curve $(u(\mathcal{P}, \Theta_i), v(\mathcal{P}, \Theta_i))$ to a set of 2D points $\{(u_1, v_1), \ldots, (u_N, v_N)\}$. For such an estimate we have the following result:

Proposition 3.2. Assuming that

– every data point (u_i, v_i) is known with the same precision, for both horizontal and vertical coordinates; all errors are independent;
– the quantities $\{\Theta_1, \ldots, \Theta_N\}$ are exactly known;
– the data points are close to the fitted curve;
– covariances are expected to be small,

then the inverse of the covariance matrix $\Lambda_{\mathcal{P}}$ (information matrix) for the optimal estimate $\mathcal{P}^*$ is

$$
\Lambda_{\mathcal{P}}^{-1} \simeq \frac{1}{\sigma^2} \left(\sum_{i=1}^{N} \frac{\partial u(\mathcal{P}^*, \Theta_i)}{\partial \mathcal{P}} \cdot \frac{\partial u(\mathcal{P}^*, \Theta_i)}{\partial \mathcal{P}}^{\mathrm{T}} \right.
$$
$$
\left. + \frac{\partial v(\mathcal{P}^*, \Theta_i)}{\partial \mathcal{P}} \cdot \frac{\partial v(\mathcal{P}^*, \Theta_i)}{\partial \mathcal{P}}^{\mathrm{T}} \right).
$$

Proof. The quantity $\mathcal{P}^*$ is a solution of the normal equations of the ξ^2 criterion. Let us write $\frac{1}{2} \frac{\partial \xi^2}{\partial \mathcal{P}} = f(\ldots)$:

$$
0 = f(\mathcal{P}^*, \{(u_1, v_1, \Theta_1), \ldots, (u_N, v_N, \Theta_N)\}) = \frac{1}{2} \frac{\partial \xi^2}{\partial \mathcal{P}}
$$
$$
= \sum_{i=1}^{N} [u(\mathcal{P}^*, \Theta_i) - u_i] \frac{\partial u(\mathcal{P}^*, \Theta_i)}{\partial \mathcal{P}} + [v(\mathcal{P}, \Theta_i) - v_i] \frac{\partial v(\mathcal{P}^*, \Theta_i)}{\partial \mathcal{P}} .
$$

Because covariances are expected to be small, it is realistic to try to estimate them using first order expansions. Then, considering the $f(\ldots)$ as a function of $\{(u_1, v_1, \Theta_1), \ldots, (u_N, v_N, \Theta_N)\}$ we have

$$\Lambda_f = \sum_{i=1}^{N} \frac{\partial f(\ldots)}{\partial(u_i, v_i)^{\mathrm{T}}} \cdot \Lambda_i \cdot \frac{\partial f(\ldots)}{\partial(u_i, v_i)^{\mathrm{T}}}^{\mathrm{T}} + \frac{\partial f(\ldots)}{\partial \Theta_i} V_{\Theta_i} \frac{\partial f(\ldots)}{\partial \Theta_i}^{\mathrm{T}}$$

$$\simeq \sigma^2 \left[\sum_{i=1}^{N} \frac{\partial u(\mathcal{P}^*, \Theta_i)}{\partial \mathcal{P}} \cdot \frac{\partial u(\mathcal{P}^*, \Theta_i)}{\partial \mathcal{P}}^{\mathrm{T}} + \frac{\partial v(\mathcal{P}^*, \Theta_i)}{\partial \mathcal{P}} \cdot \frac{\partial v(\mathcal{P}^*, \Theta_i)}{\partial \mathcal{P}}^{\mathrm{T}} \right]$$

since

$$\frac{\partial f(\ldots)}{\partial u_i} = -\frac{\partial u(\mathcal{P}, \Theta_i)}{\partial \mathcal{P}}, \quad \frac{\partial f(\ldots)}{\partial v_i} = -\frac{\partial v(\mathcal{P}, \Theta_i)}{\partial \mathcal{P}},$$

$$\Lambda_i = \begin{pmatrix} \sigma^2 & 0 \\ 0 & \sigma^2 \end{pmatrix}, \quad \text{and} \quad V_{\Theta_i} \simeq 0.$$

Considering the quantity $\mathcal{P}^*$ as given by the implicit equation $0 = f(\mathcal{P}^*, \ldots)$ we have, from the implicit function theorem:

$$\Lambda_{\mathcal{P}}^{-1} = \frac{\partial f(\ldots)}{\partial \mathcal{P}}^{\mathrm{T}} \cdot \Lambda_f^{-1} \cdot \frac{\partial f(\ldots)}{\partial \mathcal{P}}.$$

But we have:

$$\frac{\partial f(\ldots)}{\partial \mathcal{P}} = \sum_{i=1}^{N} [u(\mathcal{P}^*, \Theta_i) - u_i] \frac{\partial^2 u(\mathcal{P}^*, \Theta_i)}{\partial \mathcal{P}^2}$$

$$+ \frac{\partial u(\mathcal{P}^*, \Theta_i)}{\partial \mathcal{P}} \cdot \frac{\partial u(\mathcal{P}^*, \Theta_i)}{\partial \mathcal{P}}^{\mathrm{T}} + [v(\mathcal{P}, \Theta_i) - v_i] \frac{\partial^2 v(\mathcal{P}^*, \Theta_i)}{\partial \mathcal{P}^2}$$

$$+ \frac{\partial v(\mathcal{P}^*, \Theta_i)}{\partial \mathcal{P}} \cdot \frac{\partial v(\mathcal{P}^*, \Theta_i)}{\partial \mathcal{P}}^{\mathrm{T}}$$

$$\simeq \frac{\partial u(\mathcal{P}^*, \Theta_i)}{\partial \mathcal{P}} \cdot \frac{\partial u(\mathcal{P}^*, \Theta_i)}{\partial \mathcal{P}}^{\mathrm{T}} + \frac{\partial v(\mathcal{P}^*, \Theta_i)}{\partial \mathcal{P}} \cdot \frac{\partial v(\mathcal{P}^*, \Theta_i)}{\partial \mathcal{P}}^{\mathrm{T}}.$$

Note that, since the data points are expected to be closed to the fitted curve, second-order derivatives are not taken into account.

Combining the three last equations for $\Lambda_{\mathcal{P}}^{-1}$, $\frac{\partial f(\ldots)}{\partial \mathcal{P}}$, and Λ_f and performing a small amount of algebra yields the expected formula. $\square$

We thus are in a situation where the information matrix (inverse of the covariance) is mainly related to the Jacobians of the curve model, thus to the numerical conditioning of the equations.

In our case, considering a trajectory of $N + 1$ points given for $\Theta_i \in \{-\frac{N}{2}\Delta_\Theta, \ldots, \frac{N}{2}\Delta_\Theta\}$, and neglecting high order terms, we can use the previous result to derive explicit formulas:

$$\mathrm{Cov}(\mathcal{O}) = \mathrm{Cov}(\mathcal{C}) = \sigma^2 \begin{pmatrix} \dfrac{12}{\Delta_\Theta^2 N^3} & 0 & 0 \\ 0 & \dfrac{12}{\Delta_\Theta^2 N^3} & 0 \\ 0 & 0 & \dfrac{180(u_{i0}^2 + v_{i0}^2)}{\Delta_\Theta^4 N^5 l^2} \end{pmatrix},$$

$$\mathrm{Corr}(\mathcal{C}, \mathcal{S}) = \sigma^2 \begin{pmatrix} 0 & 0 & 0 \\ 0 & 0 & 0 \\ \dfrac{-15v_{i0}}{\Delta_\Theta^2 N^3 l} & \dfrac{15u_{i0}}{\Delta_\Theta^2 N^3 l} & \dfrac{360(u_{i1}u_{i0} + v_{i1}v_{i0})}{\Delta_\Theta^4 N^5 l^2} \end{pmatrix},$$

$$\mathrm{Cov}(\mathcal{S}) = \sigma^2 \begin{pmatrix} \dfrac{9}{4N} & 0 & \dfrac{-30v_{i1}}{\Delta_\Theta^2 N^3 l} \\ 0 & \dfrac{9}{4N} & \dfrac{30u_{i1}}{\Delta_\Theta^2 N^3 l} \\ \dfrac{-30v_{i1}}{\Delta_\Theta^2 N^3 l} & \dfrac{30u_{i1}}{\Delta_\Theta^2 N^3 l} & \dfrac{720(u_{i1}^2 + v_{i1}^2)}{\Delta_\Theta^4 N^5 l^2} \end{pmatrix}$$

with

$$l = v_{i1} u_{i0} - u_{i1} v_{i0} .$$

As expected, the higher the number of points taken into account and the higher the amplitude of the angular step Δ_Θ, the smaller the covariances. Note that for one configuration (when $l = 0$) these matrices seem to be singular. This corresponds to a non-generic situation in which the vector to the center of the curve is parallel to its tangent. In fact, the estimate is not singular but the covariances have to be computed taking higher order terms into account.

3.5 Computing Calibration Parameters

Knowing $\mathcal{P}$, we must now analyse how to recover the calibration parameters, from (3.5). In fact, we must find a relation between $\mathcal{C}$, $\mathcal{S}$ and $\mathcal{O}$ and A which is not dependent from M, C and $\boldsymbol{u}$. This is given by the following result:

Proposition 3.3. If M does not belong to the rotation axis Δ, the triplet of vectors $(\boldsymbol{u}, A^{-1} \cdot \mathcal{C}, A^{-1} \cdot \mathcal{S})$ form a direct orthogonal reference frame, and we have $||A^{-1} \cdot \mathcal{C}|| = ||A^{-1} \cdot \mathcal{S}||$.

Moreover, the following relation holds:

$$z^{\mathrm{T}} \cdot (A^{-1} \cdot \mathcal{C} + A^{-1} \cdot \mathcal{O}) = O3 + C3 = 1 .$$

Proof. Compute $(\boldsymbol{u}^{\mathrm{T}} \cdot A^{-1} \cdot \mathcal{C}) = 0$ in the third equation of (3.5), thus $\boldsymbol{u} \perp A^{-1} \cdot \mathcal{C}$.

Compute $\boldsymbol{u} \wedge A^{-1} \cdot \mathcal{C} = A^{-1} \cdot \mathcal{S}$ using (3.5).

From this result we obtain that $\boldsymbol{u} \perp A^{-1} \cdot \mathcal{S}$, and $A^{-1} \cdot \mathcal{C} \perp A^{-1} \cdot \mathcal{S}$, but also that $(\boldsymbol{u}, A^{-1} \cdot \mathcal{C}, A^{-1} \cdot \mathcal{S})$ form, if none of these vectors is null, a direct frame of reference.

Now since $\boldsymbol{u} \perp A^{-1} \cdot \mathcal{C}$, we derive from the previous equation: $||\boldsymbol{u}|| \, ||A^{-1} \cdot \mathcal{C}|| = ||A^{-1} \cdot \mathcal{S}||$, but $\boldsymbol{u}$ being unitary we obtain $||A^{-1} \cdot \mathcal{C}|| = ||A^{-1} \cdot \mathcal{S}||$.

Then computing, for instance, $||A^{-1} \cdot \mathcal{S}||^2$ we get

$$||A^{-1} \cdot \boldsymbol{S}|| = \frac{1}{Z}\sqrt{||M - C||^2 - (\boldsymbol{u}^{\mathrm{T}} \cdot M)^2} \,.$$

Finally $||A^{-1} \cdot \boldsymbol{S}|| = 0$ if and only if $||M - C|| = (\boldsymbol{u}^{\mathrm{T}} \cdot M)$, since $Z \neq 0$ is a projective factor. But $(\boldsymbol{u}^{\mathrm{T}} \cdot M) = [\boldsymbol{u}^{\mathrm{T}} \cdot (M - C)] = ||M - C|| \cos(\boldsymbol{u}, \hat{C}M)$ since $\boldsymbol{u} \perp C$ and $\boldsymbol{u}$ is unitary. We thus have $||M - C|| = (\boldsymbol{u}^{\mathrm{T}} \cdot M)$ if and only if $\boldsymbol{u}||CM$ that is if M belongs to the rotation axis Δ. If not $||A^{-1} \cdot \boldsymbol{C}|| = ||A^{-1} \cdot \boldsymbol{S}|| \neq 0$, and the triplet of vectors form a true frame of reference.

On the contrary, if M belongs to the axis of rotation, M is invariant, and neither $\boldsymbol{S}$ nor $\boldsymbol{C}$ are even defined since the trajectory reduces to a point.

The additional relation can be derived straightforward.   □

It is often not realistic to assume that M belongs to the rotation axis, because in the case of a robotic head, this axis is behind the camera. This situation could however happen for a "arm–eye" device, the camera being mounted at the end of robotic arm.

We now can rewrite and simplify (3.5) as

$$
\begin{aligned}
A^{-1} \cdot \boldsymbol{S} &= \boldsymbol{u} \wedge A^{-1} \cdot \boldsymbol{C} \,, \\
0 &= C^{\mathrm{T}} \cdot \left[\boldsymbol{u} \wedge A^{-1} \cdot \boldsymbol{O} \right] \,, \\
A^{-1} \cdot \boldsymbol{C} + A^{-1} \cdot \boldsymbol{O} &= \tfrac{1}{Z} M
\end{aligned}
\tag{3.9}
$$

which is obviously equivalent to the original equations, knowing Prop. 3.3.

The equations being put in this last form, one may see that it is easy to recover the extrinsic parameters. But more than that, these derivations have several interesting implications:

3.5.1 Equations for the Intrinsic Calibration Parameters

It is possible to generate, knowing $\mathcal{P}$, two equations related to the intrinsic parameters $(u_0, v_0, \alpha_u, \alpha_v)$ by simply considering that $A^{-1} \cdot \boldsymbol{C}$ and $A^{-1} \cdot \boldsymbol{S}$ are orthogonal and have the same magnitude:

$$
\begin{aligned}
0 = e_1 &= (\boldsymbol{C}^{\mathrm{T}} \cdot A^{-1\,\mathrm{T}} \cdot A^{-1} \cdot \boldsymbol{S}) \\
&= \tfrac{1}{\alpha_u^2}(C1 - u_0 C3)(S1 - u_0 S3) \\
&\quad + \tfrac{1}{\alpha_v^2}(C2 - v_0 C3)(S2 - v_0 S3) + C3S3 \,, \\
0 = e_2 &= (\boldsymbol{C}^{\mathrm{T}} \cdot A^{-1\,\mathrm{T}} \cdot A^{-1} \cdot \boldsymbol{C}) - (\boldsymbol{S}^{\mathrm{T}} \cdot A^{-1\,\mathrm{T}} \cdot A^{-1} \cdot \boldsymbol{S}) \\
&= \tfrac{1}{\alpha_u^2}\left[(C1 - u_0 C3)^2 - (S1 - u_0 S3)^2\right] \\
&\quad + \tfrac{1}{\alpha_v^2}\left[(C2 - v_0 C3)^2 - (S2 - v_0 S3)^2\right] + C3^2 - S3^2 \,.
\end{aligned}
\tag{3.10}
$$

Therefore, these two equations relate A with the trajectory parameters of one point, tracked in at least four views. For each point we generate two equations, therefore at least two correspondences are required.

These equations are quadratic with respect to (u_0, v_0) and linear with respect to $\left(1/\alpha_u^2, 1/\alpha_v^2\right)$. If generic, two sets of such equations are sufficient to estimate the intrinsic parameters. Moreover we have $\alpha_u > 0$ and $\alpha_v > 0$, these equations being thus unambiguous with respect to α_u and α_v.

However, expanding these equations, one can easily verify that they are linear with respect to

$$\boldsymbol{b} = (\frac{1}{\alpha_u^2}, \frac{1}{\alpha_v^2}, \frac{u_0}{\alpha_u^2}, \frac{v_0}{\alpha_v^2}, \frac{u_0^2}{\alpha_u^2} + \frac{v_0^2}{\alpha_v^2})^{\mathrm{T}} .$$

Therefore, by a suitable change of variables, it is possible to obtain linear equations related to the intrinsic parameters: considering the five components of $\boldsymbol{b}$ as unknowns the measurement equations are now linear, and it is straightforward to calculate the intrinsic parameters from an estimate of $\boldsymbol{b}$.

Thus two methods are available, one is a nonlinear method, and the equations being quadratic in u_0 and v_0, it is expected to obtain several solutions, but we might choose the right solution by considering a reasonable initial estimate for u_0 and v_0, and looking for a local solution close to the original values. The other is linear but requires at least 3 correspondences in 4 views, because we have 5 unknowns.

The covariance related to these equations can be estimated from the knowledge of $\Lambda_{\mathcal{P}}$ and using formulas similar to those used in the proof of Prop. 3.2. We obtain

$$\Lambda_{(e_1,e_2)^{\mathrm{T}}} = \frac{1}{\Delta_\Theta^4 N^5 l^2} \begin{pmatrix} a & b \\ b & c \end{pmatrix}$$

with

$$\begin{cases} a &= \frac{1}{4}\left(720\, l_0^2\, C3^2 + 2880\, l_1^2\, S3^2 + 2880\, m\, C3\, S3\right) \\ b &= \frac{1}{2}\left[1440\, m\, (C3^2 - S3^2) + (2880\, l_1^2 - 720\, l_0^2)\, C3\, S3\right] \\ c &= 2880\, l_1^2\, C3^2 + 720\, l_0^2\, S3^2 - 2880\, m\, C3\, S3 \\ l_0^2 &= u_{i0}^2 + v_{i0}^2 \\ l_1^2 &= u_{i1}^2 + v_{i1}^2 \\ l &= v_{i1}\, u_{i0} - u_{i1}\, v_{i0} \\ m &= u_{i1}\, u_{i0} + v_{i1}\, v_{i0} \end{cases} .$$

Considering P trajectories, the intrinsic parameters can be estimated efficiently using the following $\varXi^2$ criterion:

$$(u_0, v_0, \alpha_u, \alpha_v) = \mathrm{argmin}_{\{u_0, v_0, \alpha_u, \alpha_v\}} \sum_{j=i}^{P} (e_1^j, e_2^j)^{\mathrm{T}} \cdot \Lambda_{(e_1^j, e_2^j)^{\mathrm{T}}}^{-1} \cdot (e_1^j, e_2^j) .$$

The final covariance is obtained (see Prop. 3.2) as

$$\Lambda_{(u_0, v_0, \alpha_u, \alpha_v)^{\mathrm{T}}}^{-1} = \sum_{j=i}^{P} \frac{\partial(e_1^j, e_2^j)^{\mathrm{T}}}{\partial(u_0, v_0, \alpha_u, \alpha_v)^{\mathrm{T}}}^{\mathrm{T}} \cdot \Lambda_{(e_1^j, e_2^j)^{\mathrm{T}}}^{-1} \cdot \frac{\partial(e_1^j, e_2^j)^{\mathrm{T}}}{\partial(u_0, v_0, \alpha_u, \alpha_v)^{\mathrm{T}}} .$$

Knowing $\mathcal{P}$, with u_0, v_0, α_u, and α_v being estimated, we can evaluate $A^{-1} \cdot \mathcal{O}$, $A^{-1} \cdot \mathcal{S}$ and $A^{-1} \cdot \mathcal{C}$. However, (3.9) is true only if Prop. 3.3 holds, that is $A^{-1} \cdot \mathcal{C} \perp A^{-1} \cdot \mathcal{S}$ and $\|A^{-1} \cdot \mathcal{C}\| = \|A^{-1} \cdot \mathcal{S}\|$. Since this is not

necessarily the case in practice, due to numerical errors and noise, we have
to enforce these conditions to be true. This can be realized easily by rotating
symmetrically $A^{-1} \cdot C$ and $A^{-1} \cdot S$ around an axis orthogonal to both vectors
and with the minimum angle which transforms those vectors to perpendicular
vectors, while their common magnitude is set to the mean of their magnitude.
We leave the reader verify that among all transformations which generate
vectors verifying Prop. 3.3, this one minimizes the distance between $A^{-1} \cdot C$
and its transform, as it does for $A^{-1} \cdot S$.

3.5.2 Extrinsic Parameters Computation

Considering (3.9) and knowing the intrinsic parameters, we can estimate the
extrinsic parameters and the point 3D location, up to a scale factor.

We can estimate $u = \left(A^{-1} \cdot C \wedge A^{-1} \cdot S\right) / \left(\|A^{-1} \cdot S\|^2\right)$. It is equivalent
to say that each trajectory provides two linear homogeneous equations for u:

$$
\begin{aligned}
0 = e_3 &= u^{\mathrm{T}} \cdot A^{-1} \cdot C , \\
0 = e_4 &= u^{\mathrm{T}} \cdot A^{-1} \cdot S
\end{aligned}
\tag{3.11}
$$

sufficient to estimate the direction of u.

The center of rotation C can be recovered up to a scale factor. We obtain,
from (3.9)

$$
\begin{aligned}
OC = \lambda\Big[&\left(S^{\mathrm{T}} \cdot A^{-1\,\mathrm{T}} \cdot A^{-1} \cdot O\right) A^{-1} \cdot S \\
&+ \left(C^{\mathrm{T}} \cdot A^{-1\,\mathrm{T}} \cdot A^{-1} \cdot O\right) A^{-1} \cdot C\Big] .
\end{aligned}
$$

This scale factor undetermination is not a surprise, since we use monocular
cues. If the depth of a point, or the absolute distance between two points is
known, the absolute value of C can be recovered, since we have

$$
\lambda = \frac{Z}{\|A^{-1} \cdot S\|^2} .
$$

It is equivalent saying that each trajectory provides two linear homogeneous equations for C:

$$
\begin{aligned}
0 = e_5 &= C^{\mathrm{T}} \cdot \Big[\left(C^{\mathrm{T}} \cdot A^{-1\,\mathrm{T}} \cdot A^{-1} \cdot O\right) A^{-1} \cdot S \\
&\qquad - \left(S^{\mathrm{T}} \cdot A^{-1\,\mathrm{T}} \cdot A^{-1} \cdot O\right) A^{-1} \cdot C\Big] , \\
0 = e_6 &= C^{\mathrm{T}} \cdot \left(A^{-1} \cdot C \wedge A^{-1} \cdot S\right) ,
\end{aligned}
\tag{3.12}
$$

sufficient to estimate the orientation of C in the plane normal to u.

It is possible to recover the absolute position of the point in space as

$$
OM = \frac{\|OC\|\,\|A^{-1} \cdot S\| \cdot \left(A^{-1} \cdot C + A^{-1} \cdot O\right)}{\sqrt{\left(S^{\mathrm{T}} \cdot A^{-1\,\mathrm{T}} \cdot A^{-1} \cdot O\right)^2 + \left(C^{\mathrm{T}} \cdot A^{-1\,\mathrm{T}} \cdot A^{-1} \cdot O\right)^2}} .
\tag{3.13}
$$

This estimate is related to the projection of $\mathcal{O}$ in the plane orthogonal to $\boldsymbol{u}$. Using this expression will insure Z to be defined as soon as $\mathcal{O}$ is not aligned with $\boldsymbol{u}$, which corresponds again to the fact that M does not belong to the rotation axis Δ.

3.5.3 Calibration Algorithm

All the previous developments can be summarized as follows:

Proposition 3.4. In the case of the tracking of stationary targets during a fixed axis rotation for which relative angles are known, each trajectory provides two equations about the intrinsic parameters.

Moreover, knowing the intrinsic parameters, each trajectory provides an estimate of the extrinsic parameters and of the 3D location of the target.

In fact, this is all what we can compute from the coefficients of the trajectory, since $\mathcal{P}$ is of dimension 8 (9 components up to a scale factor) and yields 2 equations for the intrinsic parameters, 2 equations to compute $\boldsymbol{u}$, 1 equation to compute C with respect to $\boldsymbol{u}$, 3 equations to compute M, all these pieces of information being independent. There is thus no possibility to obtain new information from $\mathcal{P}$.

This yields the following algorithm:

---

(1) Track at least two points during a fixed axis rotation of the head, record angle and trajectory coordinates.
(2) Identify the coefficients $\mathcal{P}$ by first computing initial values from the fitting of a polynomial model using (3.8), and then refine this estimate minimizing the criterion given in (3.7).
(3) Compute the intrinsic parameters using (3.10) on at least two trajectories, and the related covariances.
(4) Compute the true values of $A^{-1}\mathcal{O}$, $A^{-1}\mathcal{S}$ and $A^{-1}\mathcal{C}$. Then rotate symmetrically $A^{-1}\mathcal{S}$ and $A^{-1}\mathcal{C}$ around $A^{-1}\mathcal{S} \wedge A^{-1}\mathcal{C}$ to obtain $A^{-1}\cdot\boldsymbol{C} \perp A^{-1}\cdot\boldsymbol{S}$, and modify symmetrically the magnitudes of $A^{-1}\mathcal{S}$ and $A^{-1}\mathcal{C}$ to get $\|A^{-1}\cdot\boldsymbol{C}\| = \|A^{-1}\cdot\boldsymbol{S}\|$.
(5) Compute the extrinsic parameters $(\boldsymbol{u}, \boldsymbol{OC}/\|\boldsymbol{OC}\|)$ and $\boldsymbol{OM}/\|\boldsymbol{OC}\|$ up to a scale factor, using (3.11), (3.12), (3.13). Compute the related covariances.

The Auto-calibration Algorithm.

The system is calibrated.

3.6 Experimental Results

3.6.1 How Stable are Calibration Parameters When Zooming?

Before evaluating our method let us verify that there is a real need to re-calibrate a visual sensor during an active visual task. In order to establish this point, we have experimented with a high quality lens, and have calibrated it for different adjustments of the zoom.

We used a well established standard method of calibration described in [3.3].

Using a Canon J8 × 6B48 high quality objective with a focal length from 6 mm to 48 mm we obtained the following numerical values:

Focal	u_0	v_0	α_u	α_v	α_u/α_v
06	262.66	248.05	128.63	190.10	0.6766
08	256.33	249.07	126.85	189.60	0.6690
09	253.77	241.16	462.17	660.38	0.6998
10	258.86	251.62	556.26	793.67	0.7009
12	252.85	242.48	668.74	953.09	0.7016
14	251.40	242.41	783.06	1114.55	0.7026
16	252.73	265.28	868.10	1235.35	0.7027
25	244.85	291.51	1427.20	2033.12	0.7020
30	247.81	299.21	1742.43	2482.23	0.7018

These values are stable for a given focal length, thus the variations are not due to the calibration method but to the optical characteristics of the lens. It is obvious from this set of data that modifying the focal length of a lens when zooming affects all intrinsic parameters of the visual sensor. In particular the optical center seems to be shifted during a zoom.

However the α_u/α_v ratio is rather stable for a reasonable range of focal length (10 to 30 mm) since it is only related to the geometrical characteristics of the pixels and the timings of the data acquisition module, but is modified for short focal length (wide angles of view) because for such a configuration the distortion at the periphery of the retina is no longer negligible.

The retinal location of the optical center (u_0, v_0) is almost linearly translated onto the retina during a zooming. This is possibly due to errors in alignment between the lens and the rotation axis used by the zoom.

Anyway, it is worthwhile to attempt to re-calibrate the system using our method.

3.6.2 Experiment 1: Parameter Estimation with Synthetic Data

We have studied the robustness of the method, considering noisy trajectories generated artificially. Gaussian noise has been added to the 2D coordinates of the points of the trajectory, the standard deviation being given in pixels.
 Computations have been done using only two trajectories.
 We have obtained the following results:

Noise [pixel]	ϵ_{u_0} [pixel]	ϵ_{v_0} [pixel]	ϵ_{α_u} [%]	ϵ_{α_v} [%]	ϵ_u [deg]	ϵ_{OC} [deg]
0	$\simeq 0$	0.11	$\simeq 0$	0.04	0.0004	0.006
0.05	0.02	4.06	0.08	6.1	0.02	0.03
0.2	6.0	11.2	11	16	1.86	7.04
0.5	3.0	8.1	10	6.7	2.12	6.25
1	4.2	6	14	25	6.12	13.23
2	12.3	11.2	23	32	12.72	22.94
5	23.2	17.1	huge	huge	none	none

Errors for (u_0, v_0) correspond to errors in the estimate of the projection of the optical center in pixel (p), errors for (α_u, α_v) are given in percentage, errors for the orientation of u and OC are given in degree.
 These results lead to three comments: (1) because the rotation axis was vertical, variations were almost in the horizontal plane, and parameters related to horizontal positions have a better estimate than parameters estimated related to vertical positions; (2) the method tolerates an error of about 2 pixels for each point of trajectory, and diverges for higher levels of noise; (3) these results have the same order of magnitudes as those obtained for standard methods of calibration [3.3–5].
 Note that this has been estimated using only two trajectories. We have repeated the same experiment but with ten trajectories and obtained, under the same conditions:

Noise [pixel]	ϵ_{u_0} [pixel]	ϵ_{v_0} [pixel]	ϵ_{α_u} [%]	ϵ_{α_v} [%]	ϵ_u [deg]	ϵ_{OC} [deg]
1	1.3	3.3	4	8	1.27	7.4

It then seems possible to increase the precision of the method, by tracking several trajectories at the same time.

3.6.3 Experiment 2: Trajectory Identification Using Real Data

We then have studied the accuracy of the mechanism of trajectory identification.

Point Tracking and Trajectory Identification. In this experiment we have tracked the center of gravity of a white blob of about 5 pixels diameter over a dark background the intensity being segmented using a statistical threshold based on the analysis of the intensity distribution. Blobs were identified detecting connected regions of bright pixels and computing the center of these points. More than 20 positions are recorded for each trajectory. This part of the experiment is entirely automatic.

The trajectory is estimated from a polynomial model computed up to order 10. The order of the development is automatically estimated using a Ξ^2 test on the residual error, the evaluation being stopped for a probability of distribution similarity better than $p = 0.99$. A typical trajectory is shown in Fig. 3.2 and Fig. 3.3. We plotted the x and y coordinates against the angle of rotation Θ, the continuous line corresponds to the true trajectory and the dashed line to the estimated trajectory. Since the trajectory was

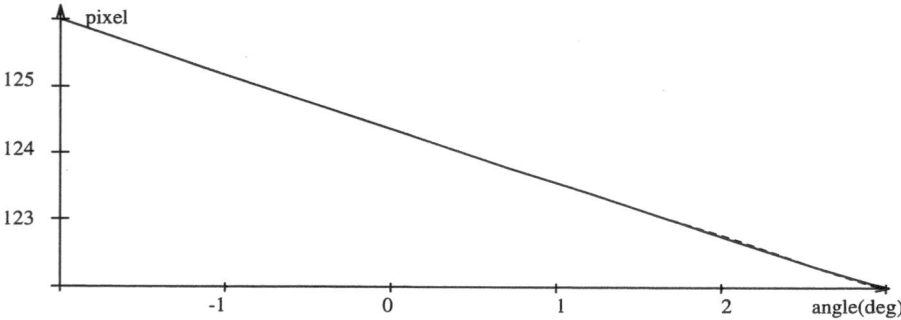

Fig. 3.2. True (*continuous line*) and estimated (*dashed line*) horizontal coordinates during a tracking, vertical scale is in pixel

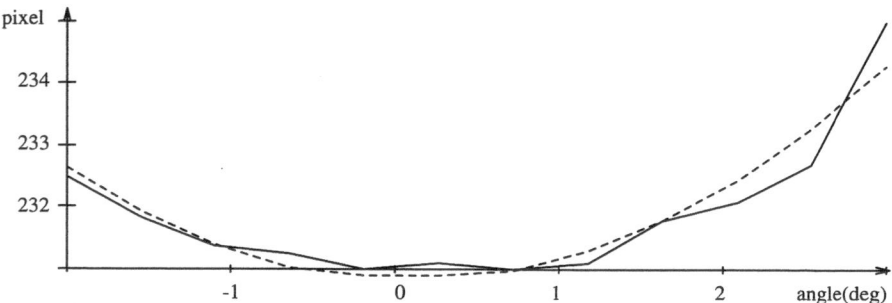

Fig. 3.3. True (*continuous line*) and estimated (*dashed line*) vertical coordinates during a tracking, vertical scale is in pixel

almost horizontal, the x estimate is almost a line and the order of the Taylor expansion has been automatically stopped after order 1. For the vertical coordinates, the development has been carried out up to the fourth order. We increased the scale for the vertical coordinate to show the differences between the two curves.

Behavior of the Algorithm. Identifying the trajectory using a polynomial model (3.8) yields an average retinal error of 2–3 pixels between the data points and the trajectory, while recomputing this estimate minimizing the proposed criterion (3.7) yields an average retinal error under 1 pixel. However minimizing the proposed criterion without a good initial estimate yields unstable results. It is thus important to perform the estimate as proposed here.

Precision of the Method. We obtained an overall precision of 0.2 pixel (standard deviations), and the order of magnitude of this value is fairly stable. This indeed means that, it is possible to obtain subpixel accuracy, for the trajectory since we average several data points.

3.6.4 Experiment 3: Parameters Estimation Using Real Data

In this experiment we tracked the corners of the calibration grid shown in Fig. 3.4 and identified the targets using a standard calibration method as described in [3.11]. This part of the experiment is entirely automatic.

The parametric model of all trajectories has been represented in Fig. 3.5. We observed 128 points during 31 views. It is visible that there are false matches for six of the trajectories, but we did not correct it, assuming the method will be robust enough to deal with this bias.

Intrinsic Parameters: Comparison with a Standard Method. Unfortunately, it is not possible to know the "true" value of the intrinsic parameters of a visual sensor. We thus only compared the results obtained with our new method with respect to the results obtained using a standard method [3.3],

Fig. 3.4. Using a calibration grid to compare our method with a well established method of calibration

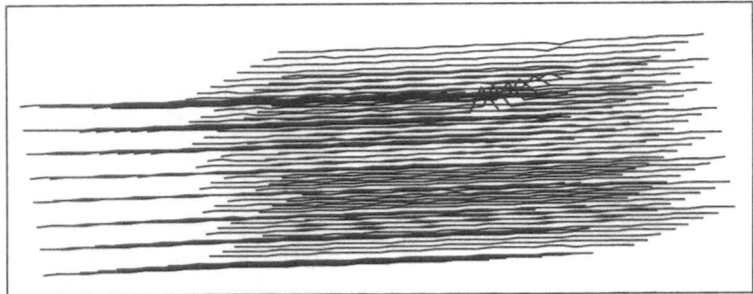

Fig. 3.5. Parametric trajectories identified during the tracking of 128 points along 30 frames

under similar conditions. The main difference being, of course, that we do not introduce any information about the grid in our software.

Using the standard method several times we obtained the following results:

	u_0	v_0	α_u	α_v
	240.4	215.6	825.5	1228.7
	241.8	226.5	826.2	1257.5
	255.1	223.6	830.9	1215.3
	242.4	215.4	848.1	1262.1
	247.8	213.3	845.5	1227.1
	238.5	224.0	850.3	1170.9
	239.4	205.5	837.1	1223.7
	247.7	244.0	859.8	1222.6
	250.8	224.8	820.6	1240.8
	255.2	236.5	838.4	1234.6
	236.6	203.1	835.4	1214.6
mean	245.06	221.12	837.98	1227.08

Since such a method does not provide a covariance for the estimate we roughly have estimated the standard deviations from a set of 20 measurements. With such a method the location of the optical center is known with a precision of 2–3 pixels, and the scale factors with a precision of 5 %.

Using the method developed in this book we obtained the following performances for the same run:

```
-u0 245.93 SD=0.03 -v0 236.22 SD=0.3
-au 832.06 SD=0.5 -av 1280.99 SD=30.0 .
```

We have controlled these numbers running the experiment several times, and obtained very similar values. We can state two fundamental facts: (1) our method yields better results, in term of variances, than the standard method because we do not analyse only one image of the grid but a full set; (2) the values have a better accuracy for parameters related to horizontal positions than those related to vertical positions (u_0 and α_u are much more accurate than v_0 and α_v) because the motion of the head was done almost in a horizontal plane as visible on Fig. 3.5: on one hand horizontal related values are close to the average value obtained with the standard method while vertical related values are not, on the other hand estimated covariances are higher for vertical related values. In any case, the covariances seem to correspond to the true precision of the data, and an iterative statistical filter might manage the fact that we only have a partial observation of the parameters.

Since these results are limited because the other calibration method is itself imprecise, we have indirectly checked our results verifying if, after the intrinsic calibration, the constraints of Prop. 3.3 are verified. They should if the calibration is perfect. Computing

$$ \mathrm{Err}_n = \sqrt{\frac{1}{N} \sum_{i=1}^{N} \left(\frac{\|A^{-1} \cdot \boldsymbol{C}\| - \|A^{-1} \cdot \boldsymbol{S}\|}{\|A^{-1} \cdot \boldsymbol{C}\| + \|A^{-1} \cdot \boldsymbol{S}\|} \right)^2} $$

as an indicator of the relative error in magnitude between the two vectors, and

$$ \mathrm{Err}_a = \sqrt{\frac{1}{N} \sum_{i=1}^{N} \left[\frac{\varPi}{2} - \arccos \left(\frac{A^{-1} \cdot \boldsymbol{S}^{\mathrm{T}} \cdot A^{-1} \cdot \boldsymbol{C}}{\|A^{-1} \cdot \boldsymbol{S}\|^2} \right) \right]^2} $$

as an indicator of the error in orientation between these two vectors, we obtained

```
Err n 0.0217241 Err a 3.45087 deg .
```

These values should have been zero if the two vectors had the same magnitude and were orthogonal.

Note that, in order to obtain a relevant estimate of the vertical calibration parameters, a rotation in the vertical plane is to be made.

Extrinsic Parameters. The intrinsic parameters being calibrated, the rest of the experimentation is mainly a motion and structure from motion paradigm, in which some assumptions are made about the kind of motion performed.

Extrinsic parameters are the axis orientation parameterized by the unitary vector $\boldsymbol{u}$ and the axis location up to a scale factor parameterized by the orientation α of C in the plane perpendicular to $\boldsymbol{u}$.

A typical set of results is, using a recursive mechanism of Kalman filtering to propagate the estimations and their covariances:

```
u: (0.032 -0.997 -0.072) p(u)/p0: 0.21 alpha -73.940 p(alpha)/p0: 0.34
u: (0.027 -0.998 -0.059) p(u)/p0: 0.08 alpha -80.790 p(alpha)/p0: 0.12
u: (0.029 -0.998 -0.056) p(u)/p0: 0.05 alpha -70.039 p(alpha)/p0: 0.60
u: (0.029 -0.998 -0.057) p(u)/p0: 0.05 alpha -90.655 p(alpha)/p0: 0.78
u: (0.024 -0.998 -0.053) p(u)/p0: 0.06 alpha -65.612 p(alpha)/p0: 0.89
u: (0.029 -0.998 -0.057) p(u)/p0: 0.06 alpha -90.123 p(alpha)/p0: 0.74
u: (0.026 -0.998 -0.058) p(u)/p0: 0.07 alpha -85.725 p(alpha)/p0: 0.45
u: (0.026 -0.998 -0.049) p(u)/p0: 0.04 alpha -83.133 p(alpha)/p0: 0.27
u: (0.026 -0.999 -0.036) p(u)/p0: 0.15 alpha -79.612 p(alpha)/p0: 0.04
u: (0.029 -0.999 -0.035) p(u)/p0: 0.16 alpha -88.522 p(alpha)/p0: 0.63
u: (0.032 -0.999 -0.033) p(u)/p0: 0.19 alpha -90.430 p(alpha)/p0: 0.76
```

where $p(u)$ and $p(\alpha)$ are the probabilities for the measure not be equal to the estimated mean value. We have computed the mean value of each estimate, and then rejected measures which probability to correspond to the estimate are lower than a threshold p_0 [3.26]. The final estimate is performed using plausible measures only.

We finally have obtained:

```
>> u =        (        0.03         -1       -0.051 )

              (     4.3e-05      4.6e-06    -6.2e-05 )
>> Cov(u) =   (     4.6e-06      2.7e-06    -4.6e-05 )
              (    -6.2e-05     -4.6e-05     0.00089 )

>> Alpha = -81.689154 SD=8.270093
```

while the expected values were $u = (0, -1, 0)$ and $\alpha = -90.0$, considering the mechanical configuration of the camera on the mount [3.1].

We thus have obtained coherent estimates of the extrinsic parameters, but the axis orientation u is always estimated with a high accuracy, whereas the orientation of C is a rather unstable parameter. This is expected since u only depends on the relative orientation of $A^{-1} \cdot S$ and $A^{-1} \cdot C$, whereas C is a complex, thus more sensitive, function of all parameters.

As for the computation of the intrinsic parameters, we also have measured how the constraint of Prop. 3.3 is accurate. We obtained a mean quadratic error of: Err a 1.2421 deg the order of magnitude being the same as before.

3D Reconstruction. A simple way to check if the calibration process is correct is indeed to perform a 3D reconstruction of the point in space since this is provided by our method.

A view of the reconstructed grid is shown on Fig. 3.6, the image has been tilted by an angle of 20 deg around a diagonal axis.

The quality of the result is reasonable, but this result is obtained on the points which have been used for the calibration process. The calibration parameters are thus adapted to this configuration.

Fig. 3.6. A tilded view of the reconstructed grid

3.7 Conclusion

We do not use any information about the absolute positioning of points in space, as done normally in calibration procedures [3.3], and this is the real improvement of this method. We have made use of the nature of the motion a robotic system, and obtain better results than a general projective algorithm [3.7, 9, 27] in this case, since this configuration is singular for general projective algorithms [3.19]. It seems that not many "simple" motions can be used for auto-calibration. For instance, pure translation cannot induce auto-calibration as shown in the appendix. Pure rotations can be used as it will demonstrated in a future study.

As soon as one can track at least two stationary points in at least four views, during a fixed axis rotation, the system can be calibrated. It can be used in a clustered environment, performing in parallel other tasks. Moreover, the two points are to be tracked either at the same time, or at different times but in the same conditions. Since we only look at the complete trajectory, we do not need to track synchronously these points.

Using a simple pin-hole model the intrinsic parameters is only approximative and reliable locally with respect to the retinal location or the 3D depth. Since our method can be applied in different windows of attention corresponding to objects located in a given area of the visual surroundings it is thus possible to adjust the parameters to fit to this neighbourhood.

We also demonstrate that auto-calibration of a robotic system is possible when using an active visual sensor. In this case not only the camera can be calibrated but also the mount itself, since the rotation axis is estimated in the frame of reference of the camera. This helps to calibrate the robotic head as described in [3.1].

However, this method needs to be realized in a precise experimental situation: the robotic head should only perform a fixed axis rotation but should not being subject to other motions (if mounted on a vehicle, the vehicle has to be stopped).

This method is complementary to what has been obtained using projective invariants for auto-calibration [3.7], when calibrating stereo camera geometry using a variational principle [3.9, 27] or when calibrating up to an affine transformation as done in [3.28]. In comparison with these studies, our method seems to be more restrictive since we (1) make rather strong assumptions about the kind of motion performed by the system (note that [3.9, 27] also consider a restrained rigid motion, while [3.7] considers a general rigid motion) and (2) we have to track these points during several views, while the other methods require two or three views only. However, we only need to track two points while the other method requires seven or eight points, and it seems that our experimental results are more robust. Our method is thus less general but well adapted when fixed axis rotations are used.

Our implementation corresponds to a recursive process (iterative filtering) correcting these parameters in function of the related precision of the initial estimate and eliminating false measures.

3.8 Comparison with the Case of Known Translations

On some systems, there are not only angular joins, but also linear joins performing pure translations [3.2]. In that case the rectilinear displacement is known with a good precision from the sensors. However, the direction of the axis is not known in a retinal frame of reference because the rigid transformation from the retina to the mount is usually not known.

In the case of such a fixed axis translation in the direction of t (we choose $||t|| = 1$) and with a linear displacement τ, the corresponding equations are the following:

$$\mathcal{Z}(\Theta)\,(u, v, 1)^{\mathrm{T}} = A \cdot (M + \tau t) \;\Leftrightarrow\; (u, v, 1)^{\mathrm{T}} = \frac{Z}{\mathcal{Z}(\Theta)}\,[\tau \boldsymbol{T} + \boldsymbol{\mathcal{O}}]$$

where

$$\boldsymbol{T} = \begin{pmatrix} T1 \\ T2 \\ T3 \end{pmatrix} = \frac{1}{Z} A \cdot t \quad \text{and} \quad \boldsymbol{\mathcal{O}} = \begin{pmatrix} O1 \\ O2 \\ O3 \end{pmatrix} = \frac{1}{Z} A \cdot M,$$

thus

$$u = u_0 + \frac{T1\tau + O1}{T3\tau + O3},$$

$$v = v_0 + \frac{T2\tau + O2}{T3\tau + O3}.$$

From these equations, the direction of the axis of translation can be recovered. By eliminating t we now have four independent equations only.

It is then clear that the lens cannot be calibrated from these equations, because A cannot be computed in this paradigm due to the lack of equations.

However the lens being calibrated, it is possible to eliminate Z and recover the absolute position of the point in space and the direction of the translation as

$$M = \frac{A^{-1} \cdot \mathcal{O}}{||A^{-1} \cdot \boldsymbol{T}||}, \qquad t = \frac{A^{-1} \cdot \boldsymbol{T}}{||A^{-1} \cdot \boldsymbol{T}||}.$$

Then, using linear joins leads to much simpler equations, allows to calibrate the extrinsic parameters and to recover the depth of the point, but does not allow the calibration of the lens (intrinsic parameters).

3.9 Application to the Case of a Binocular Head

In the case of an active stereo "head" we have to calibrate the intrinsic parameters for both visual sensors and the extrinsic parameters related to relative positioning of the two cameras as shown in Fig. 3.7.

On such a system, we can very easily calibrate the intrinsic parameters by combining pan and tilt fixed axis rotations: pan rotations induce horizontal retinal disparities and allow to calibrate the horizontal intrinsic calibration parameters whereas tilt rotations induce vertical retinal disparities and allow to calibrate the vertical intrinsic calibration parameters.

Moreover, the location of the rotation axis up to a scale factor is known in a retinal frame of reference, and it is therefore possible to compute the

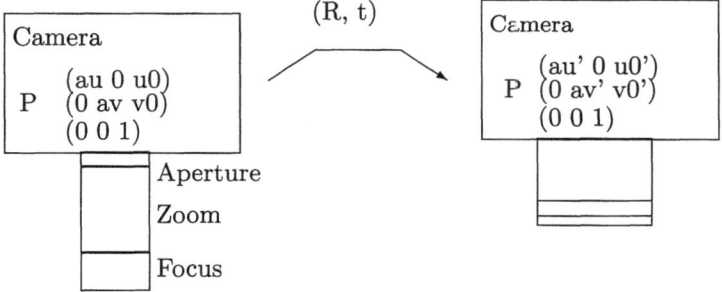

Fig. 3.7. Parameters in the calibration of a stereo head

extrinsic parameters up to a scale factor. The scale factor can be evaluated using either a standard method of calibration or more simply by measuring manually one distance in the system.

3.10 Instantaneous Equations for Calibration

In our paradigm we track a few sets of trajectories but perform the calibration "off-line", that is after the whole trajectory have been identified. The reason for this is to obtain robust estimates of the calibration parameters.

It is also possible to look for relations occurring as soon as point correspondences have been obtained between two views. This problem has been already investigated [3.8] and solved [3.7] but when no assumption about the motion can be done.

However, in our case, the rigid motion is specific and we would like to investigate whether assuming the displacement is an affine rotation around a fixed axis yields simplifications in the general calibration algorithm of Faugeras-Maybank [3.8].

We also would like to discuss why, we better use the whole tracked trajectories rather than local point correspondences.

3.10.1 Reviewing the Definition of the Fundamental Matrix

The 2D projection m of a 3D point M is defined through $Z\, m = P \cdot M$, as already explained, while between two views the 3D point undergoes a rigid transformation written $M' = R(\boldsymbol{u}, \Theta) \cdot M + \boldsymbol{t}$, but in our case we have $\boldsymbol{t} \perp \boldsymbol{u}$.

The fact that $\boldsymbol{t} \perp \boldsymbol{u}$ is equivalent to have a rotation around a fixed axis, as the reader can easily verify, since we have the relations:

$$OC = \frac{1}{2}\left[\boldsymbol{t} + \frac{\sin(\Theta)}{1 - \cos(\Theta)}\boldsymbol{u} \wedge \boldsymbol{t}\right]$$

and

$$\boldsymbol{t} = \sin(\Theta)\boldsymbol{OC} \wedge \boldsymbol{u} + (1 - \cos(\Theta))\boldsymbol{OC} \,.$$

The Longuet–Higgins relation [3.29], generalized by Maybank and Faugeras [3.7], states that $(M', R(\boldsymbol{u}, \Theta) \cdot M, \boldsymbol{t})$ are coplanar, obviously, and this yields

$$
\begin{aligned}
|P^{-1} \cdot m', R(\boldsymbol{u}, \Theta) \cdot P^{-1} \cdot m, \boldsymbol{t}| &= 0 \\
\Leftrightarrow m'^{\mathrm{T}} \cdot \underbrace{\left(P^{-1\mathrm{T}} \cdot \tilde{\boldsymbol{t}} \cdot R(\boldsymbol{u}, \Theta) \cdot P^{-1}\right)}_{F} \cdot m &= 0 \,.
\end{aligned}
$$

The matrix F is called the fundamental matrix [3.7], it has nine elements, but seven degrees of freedom, since F is known up to a scale factor and $\det(F) = 0$, as we will redemonstrate here.

This matrix F thus resumes all information about intrinsic and extrinsic parameters in the case of points correspondences.

3.10.2 Characterizing the Essential Matrix for Fixed Axis Rotations

Let us consider the matrix $E = P^{\mathrm{T}} \cdot F \cdot P = \tilde{t} \cdot R(u, \Theta)$, called the essential matrix, cf. [3.30].

We have the following fundamental characterization, in the case of a fixed axis rotation:

Proposition 3.5. A 3×3 matrix E is of the form $E = \tilde{t} \cdot R(u, \Theta)$ with $t \perp u$ if and only if:

(1) $\det(E) = 0$,
(2) $\mathrm{trace}(E) = 0$,
(3) $\det\left(E + E^{\mathrm{T}}\right) = 0$,
(4) $Q_E = \left(E + E^{\mathrm{T}}\right) \cdot \left(E - E^{\mathrm{T}}\right) \cdot \left(E + E^{\mathrm{T}}\right) = 0$.
(5) Considering the eigenvectors of $\frac{1}{2}\left(E + E^{\mathrm{T}}\right)$ written v_i and their associated eigenvalues, say λ_i, we have: $\lambda_i \, v_j^{\mathrm{T}} \cdot \frac{1}{2}\left(E - E^{\mathrm{T}}\right) \cdot v_k \geq 0$ for any $i \neq j \neq k$.

Proof. Necessity: Consider a matrix $E = \tilde{t} \cdot R(u, \Theta)$ with $t \perp u$. Using the Rodriguez formula, we have:

$$E = \cos(\Theta)\tilde{t} + \sin(\Theta)u \cdot t^{\mathrm{T}} + [1 - \cos(\Theta)](t \wedge u) \cdot u^{\mathrm{T}}.$$

In order to show this proposition the simplest way is to consider the unary eigenvectors and real eigenvalues of $\frac{1}{2}\left(E + E^{\mathrm{T}}\right)$, which are:

$$v_0 = -\sqrt{\frac{1-c}{2}}\, t\|t\| - \sqrt{\frac{1+c}{2}}\, u \wedge t\|t\|,$$

$$\lambda_0 = 0,$$

$$v_+ = +\frac{1}{\sqrt{2}}u + \frac{\sqrt{1+c}}{2}\, t\|t\| - \frac{\sqrt{1-c}}{2}\, u \wedge t\|t\|,$$

$$\lambda_+ = \|t\|\,\sqrt{\frac{1-c}{2}},$$

$$v_- = -\frac{1}{\sqrt{2}}u + \frac{\sqrt{1+c}}{2}\, t\|t\| - \frac{\sqrt{1-c}}{2}\, u \wedge t\|t\|,$$

$$\lambda_- = -\|t\|\,\sqrt{\frac{1-c}{2}}$$

with $c = \cos(\Theta)$. The expression of E in the (v_+, v_-, v_0) frame of reference is

$$
E = \|t\| \begin{pmatrix} a & 0 & -b \\ 0 & -a & b \\ b & -b & 0 \end{pmatrix}
$$

$$
= \|t\|[a\left(v_+ \cdot v_+^{\mathrm{T}} - v_- \cdot v_-^{\mathrm{T}}\right)
$$

$$+b\,(\tilde{\boldsymbol{v}}_+ + \tilde{\boldsymbol{v}}_-)]\quad \begin{cases} a &=& \sqrt{(1-c)/2} \\ b &=& \sqrt{1+c}/2 \end{cases}.$$

It is then straightforward to obtain: $\text{trace}(E) = \det(E) = \det(E + E^{\mathrm{T}}) = 0$, $Q_E = 0$, and $\lambda_i\,\boldsymbol{v}_j^{\mathrm{T}} \cdot \frac{1}{2}(E - E^{\mathrm{T}}) \cdot \boldsymbol{v}_k \geq 0$, for any $(i, j, k) \in \{0, -, +\}^3$.

Sufficiency: Considering a frame of reference in which $\frac{1}{2}\left(E + E^{\mathrm{T}}\right)$ is diagonal. We can write

$$E = \begin{pmatrix} \lambda_1 & 0 & 0 \\ 0 & \lambda_2 & 0 \\ 0 & 0 & \lambda_3 \end{pmatrix} + \begin{pmatrix} 0 & -z & y \\ z & 0 & -x \\ -y & x & 0 \end{pmatrix}$$

in this frame of reference. The first term corresponds to the symmetric part of E and the second to its skew-symmetric part. This is true for any matrix.

Now if $\det(E + E^{\mathrm{T}}) = \lambda_1\lambda_2\lambda_3 = 0$ one among these three values is zero, say λ_3. Then, if $\text{trace}(E) = \lambda_1 + \lambda_2 = 0$, we have $a = \lambda_1 = -\lambda_2$. We can always consider $a \geq 0$ since λ_1 and λ_2 are of opposite sign, thus one, say λ_1, is positive. If $a = 0$ E is skew-symmetric thus corresponding to a form $\tilde{\boldsymbol{t}} \cdot R(\boldsymbol{u}, \Theta = 0)$, and the proof is completed. Let us consider $a > 0$. We have

$$Q_E = \begin{pmatrix} 0 & 8a^2 z & 0 \\ -8a^2 z & 0 & 0 \\ 0 & 0 & 0 \end{pmatrix} = 0$$

and since $a \neq 0$, $z = 0$. Please note that the condition $Q_E = 0$ yields one equation only. Moreover, $\det(E) = a(x^2 - y^2)$ and we have $b = |x| = |y|$.

But $\lambda_1\,\boldsymbol{v}_2^{\mathrm{T}} \cdot \frac{1}{2}\left(E - E^{\mathrm{T}}\right) \cdot \boldsymbol{v}_3 = -ax$ and $\lambda_2\,\boldsymbol{v}_1^{\mathrm{T}} \cdot \frac{1}{2}\left(E - E^{\mathrm{T}}\right) \cdot \boldsymbol{v}_3 = -ay$, then $x \geq 0$ and $y \geq 0$, thus $b = x = y$. It is important to note here, although this is not obvious, that these inequalities enforce the orthogonal matrix R to be a true rotation but not the combination of a rotation and a symmetry.

Finally the matrix E is of the form

$$E = \begin{pmatrix} a & 0 & -b \\ 0 & -a & b \\ b & -b & 0 \end{pmatrix},$$

as obtained previously.

Taking $\boldsymbol{u} = (1/\sqrt{2}, -1/\sqrt{2}, 0)^{\mathrm{T}}$ and $\boldsymbol{t} = (b, b, a)^{\mathrm{T}}$ with $\Theta = \arccos[(2b - a)/(2b + a)]$ (well defined because $a > 0$ and $b \geq 0$) a straightforward computation yields: $E = \alpha\,\tilde{\boldsymbol{t}} \cdot R(\boldsymbol{u}, \Theta)$, as expected.

Please note we had to use all five constraints to recover E. They thus form a minimal set of characteristic equations.   □

In order to analyse this result, let us count the equations: E is a matrix with nine components but has only six degrees of freedom. We thus expect three relations to occur in the general case and four relations in the particular case where translation and rotation axis are orthogonal. This is the case here, since the constraint Q_E reduces to one equation, while the last constraint let us obtain a true rotation. For a general essential matrix [3.30], three relations are obtained. More precisely:

Proposition 3.6. For an essential matrix $E = \tilde{t} \cdot R(\boldsymbol{u}, \Theta)$, we have $\boldsymbol{t} \perp \boldsymbol{u}$ if and only if trace$(E) = 0$.

Proof. Let us use the following result from Maybank [3.30]. "A matrix E with det$(E) = 0$ is an essential matrix if and only if the eigenvalues of the symmetric part of E verify $\lambda_1 + \lambda_2 = \lambda_3$, and the vector $\boldsymbol{v}$ such that $\tilde{v}$ corresponds to the skew-symmetric part of E is orthogonal to the eigenvector of the symmetric part of E related to λ_3."

This means that E is of the form

$$E = \begin{pmatrix} \lambda_1 & 0 & y \\ 0 & \lambda_2 & -x \\ -y & x & \lambda_3 \end{pmatrix}$$

in the frame of reference associated with the eigenvectors of $\frac{1}{2}\left(E + E^\mathrm{T}\right)$.

Necessity: If E is an essential matrix with $\boldsymbol{t} \perp \boldsymbol{u}$ the previous proposition yields trace$(E) = 0$ as expected.

Sufficiency: Consider an essential matrix E with trace$(E) = 0$. We thus have $\lambda_1 + \lambda_2 + \lambda_3 = 0$ but $\lambda_1 + \lambda_2 = \lambda_3$, thus $\lambda_3 = 0$ and $\lambda_1 = -\lambda_2 = a$. Because det$(E) = a(x^2 - y^2)$ yielding $x = \pm y$ this matrix verifies the same conditions as in the previous proposition and is thus an essential matrix with $\boldsymbol{t} \perp \boldsymbol{u}$.

The cases where R is not a rotation but the combination of a rotation and a symmetry has been discussed previously also. $\qquad\square$

3.10.3 Calibration Using the Fundamental Matrix

The previous properties allow to envisage another method of calibration, similar to what was proposed by Faugeras et al. [3.7]: (1) look for at least 7 correspondences between two viewpoints, (2) compute the fundamental matrix F with its related constraints, (3) generate equations about the intrinsic parameters, (4) knowing P, compute the essential matrix, and the extrinsic parameters.

In the present case F is subject to two constraints:

$$0 = \det(E) = \det(P^\mathrm{T} \cdot F \cdot P) = \det(P)^2 \det(F) \quad \Rightarrow \quad 0 = \det(F) ,$$
$$0 = \det(E + E^\mathrm{T}) = \det(P)^2 \det(F + F^\mathrm{T}) \quad \Rightarrow \quad 0 = \det(F + F^\mathrm{T})$$

since $\quad E^\mathrm{T} = (P^\mathrm{T} \cdot F \cdot P)^\mathrm{T} = P^\mathrm{T} \cdot F^\mathrm{T} \cdot P$.

The fact that we have $\boldsymbol{t} \perp \boldsymbol{u}$ does not produce additional equations about the intrinsic parameters but simplifies one of them which writing

$$F = \begin{pmatrix} F11 & F12 & F13 \\ F21 & F22 & F23 \\ F31 & F32 & F33 \end{pmatrix} ,$$

is

$$
\begin{aligned}
0 = \mathrm{trace}(E) \quad = \quad & (\alpha_u^2 + u_0^2)F11 + (\alpha_v^2 + v_0^2)F22 + F33 \\
& +u_0(F31 + F13) + v_0(F32 + F23) \\
& +u_0v_0(F12 + F21)
\end{aligned}
$$

which is quadratic with respect to (u_0, v_0) and linear with respect to (α_u^2, α_v^2), as with the method developed in this chapter. This equation plus the fact that $Q_E = 0$ can be used as measurement equations to determine the intrinsic parameters, as in [3.7].

Moreover, the extrinsic parameters are simple functions of the eigenvectors and eigenvalues of E, as obtained during the proof of Prop. 3.5. In particular we have $E^{\mathrm{T}} \cdot t = 0$ to compute t up to a scale factor and $E \cdot t = ||t||^2 \sin(\Theta)u$ to obtain u with $||u|| = 1$, and $\tan(\Theta) = (u^{\mathrm{T}} \cdot Et)/[u^{\mathrm{T}} \cdot E(u \wedge t)]$ to obtain Θ.

3.10.4 Discussion

When applying the fundamental matrix method in the case of a fixed axis rotation, it is indeed possible to calibrate the system, and our specific situation (rotation around a fixed axis) seems to induce some simplifications, with respect to the general method of [3.7], but not with respect to the method proposed in this book.

In our paradigm we have just 7 unknowns: 4 intrinsic parameters (u_0, v_0, a_u, a_v), 3 extrinsic parameters (2 for u which is a unary vector and 1 for C since it is known up to a scale factor and orthogonal with u). But F provides only 6 equations, since it has 9 components, known up to a scale factor, and subject to two constraints. As in the general case, it is thus not possible to observe the unknowns with one set of correspondences (two views), but two correspondences (three views) are a minimum.

Since, with this method we need to track at least 7 points instead of 2, we observe only local, ill-conditioned, information (point correspondences) instead of global information (parametric trajectories). Because both methods yield the same kind of equations (same number of equations, same degrees in the algebraic relations), it seems to be more reasonable to implement the method described in this book than the general method of calibration reviewed here.

4. Inertial Cues in an Active Visual System

On a robot, two types of inertial information can be computed: The instantaneous **self-motion** (also called either self motion or vection), of the robot, and the **angular orientation** of the robot in space.

This information can be used for different tasks, such as navigation, or stabilization of the robot, as in classical inertial systems. However, in the case of a robotic system, other tasks appear: intrinsic and extrinsic calibration, trajectory generation, assembly tasks, grinding operations, etc. Inertial measurements could be useful in each case.

In this chapter we introduce this new cue and discuss how to calibrate an inertial system on a robot. We then demonstrate that, considering some suitable hypothesies, we can separate the gravity from linear accelerations. The consequence of this mechanism is that we can obtain an estimation of the vertical, combine it with a visual estimation of the vertical and reanalyse visual information considering this important 3D cue.

4.1 Introduction

The use of inertial measurements on a robot is suggested in several points:

Several industrial realizations use inertial systems either for the navigation of airplanes – or other flying engines – around the earth, or for the stabilization of cameras when filming in a vehicle. Measurements are very precise, computation can be performed in real time, while information about motion and orientation is provided. Therefore, robot navigation and stabilization could also take advantage of this technology.

In human beings and in animals, inertial information is provided by the vestibular system in the inner ear, and are essential for spatial orientation, equilibrium, and navigation of the animal in its surroundings. This suggests that inertial measurements are required to perform such high-level tasks.

Inertial measurements yield the same kind of information as that provided by passive navigation algorithms using artificial vision, but with a different dynamic range and precision. Thus, cooperation between these two sensory modalities may be useful for the elaboration of high-level representations.

In addition, all the processes involved in such tasks have to be based on internal representations of the robot geometry and dynamic, and on internal representations of robot and object motion and orientation. The elaboration of such representations using inertial measurements has to be considered first, and the present study focuses on this point.

4.2 The Use of Inertial Forces in a Robotic System

4.2.1 Origins of Inertial Forces on a Robot

- Displacements of the robot due to the action of its effectors (**intrinsic displacement**): the displacement is then partially known, either from the parameters of the command, or from proprioceptive information. The use of such inertial information could be either:
  - to cooperate with odometric information, to compute a better estimation of intrinsic displacement, taking the characteristics of the different sensors – in terms of dynamic range and precision – taking into account,
  - the use of redundancy between the different sources of information in order to calibrate the data from odometric and inertial devices.
- Displacements of the robot not due to its effectors (the robot is on a vehicle, or is moved by an external independent device), referred as **extrinsic displacements**. The role of inertial sensors, in cooperation with other sensors is then to estimate these displacements.
- Unwanted vibrations due to actuators, and to mechanical transmissions. Feedback systems can use inertial information to eliminate these vibrations, at low frequencies (0–100 Hz).

Origin of inertial forces

The **principle of generalized relativity of Einstein,** states that only: *the specific force on one point* and the *the angular instantaneous velocity,* but no other quantity concerning motion and orientation with respect to the rest of the universe, can be measured from physical experiments inside an isolated closed system.

Then, using inertial measurements, only **linear accelerations** and **angular velocities** can be estimated, but nothing concerning absolute linear velocity, nor absolute angular orientation, nor absolute location in space. However, on earth, the gravity force is added to the specific forces, and in a Galilean frame of reference, if r is the position vector of a point having a mass m, the fundamental law of the mechanics is

$$\frac{d^2 r}{dt^2} = \ddot{r} = \frac{F}{m} = f + g$$

where g is the gravity acceleration.

This additional particular gravitational force provides also information about the absolute vertical in an earth frame of reference.

On a robot these inertial forces can have several origins, and depending upon their origin the inertial information should have different functions.

4.2.2 Distinction with Inertial Navigation on Vehicles

The theories of inertial navigation on airplanes or other vehicles and their related implementations cannot directly be used on a robotic system, for two reasons: (1) The theories study long displacements around the earth, at high velocities, where quantities such as the earth radius, or the gravity field variations, are taken into account and are used in the models. On the contrary, the orders of magnitude of a robot displacements are quite different, and other hypotheses must be used. (2) The cost of an inertial system must be similar to that of other devices used on a robot, for example a visual sensor. Therefore, inertial system on a robot cannot be the same as the very expensive ones used on airplanes, and the choice of realistic sensors has to be made.

In fact, either in the case of a mobile robot or of a manipulator, the displacement is bounded and several realistic physical hypotheses can be made:

- Inertial forces can be calculated using the classical laws of the kinematics and dynamics.
- The gravity field is a constant, homogeneous, and isotropic field of acceleration, not varying during robot displacements.
- Astronomic movements such as earth rotation have no influence, as generators of inertial forces, on the robot.

Let us discuss the last two points: If v is the relative linear velocity of a point and ρ the relative angular velocity, in an earth frame of reference, r the vector between the earth center and the point, and Ω the angular velocity of the earth, the point absolute linear velocity and angular absolute velocity will be

$$\dot{r} = v + \Omega \wedge r ,$$
$$\omega = \rho + \Omega$$

where $\wedge$ is the cross product of two 3D vectors. Since v is a screw we write in an absolute frame of reference:

$$\frac{\delta v}{\delta t} = \dot{v} + \omega \wedge v ,$$
$$\ddot{r} = (\dot{v} + \omega \wedge v) + \Omega \wedge (\dot{r})$$
$$= \dot{v} + (\rho + 2 \cdot r) \wedge v + \Omega \wedge (\Omega \wedge r) .$$

The fundamental law will take the form of

$$\dot{v} = f + g_1 - (\rho + 2 \cdot \Omega) \wedge v ,$$
$$g_1 = g - \Omega \wedge (\Omega \wedge r) .$$

Quantitative values are $\|g\| \simeq 9.81$ m/s^2, and $\|\Omega\| \simeq 7.27 \cdot 10^{-5}$ rad/s, while $\|r\| \simeq 64 \cdot 10^{+5}$ m with a relative variation, all around the earth, of

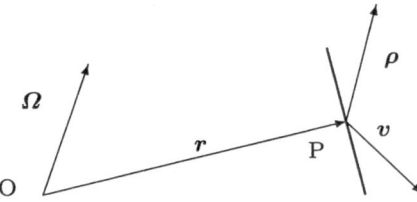

Earth Center   Earth Surface

Fig. 4.1. Representation of a movement at the earth surface

less than $4 \cdot 10^{-3}$, thus fairly constant when small displacements at the earth surface. We then have, if $\Delta \rho$ is the angular rate sensor precision:

$$\|\boldsymbol{\Omega} \wedge (\boldsymbol{\Omega} \wedge \boldsymbol{r})\| \leq 3.4 \cdot 10^{-2} \text{ m/s}^2 ,$$
$$\|\boldsymbol{\Omega}\| \ll \|\Delta \rho\|$$

while the direction of this vector is orthogonal to $\boldsymbol{r}$ aligned with $\boldsymbol{g}$, but remains constant.

It becomes obvious that the inertial forces due to the earth rotation are not to be taken into account. The vertical will be defined using $\boldsymbol{g}$ instead of $\boldsymbol{g_1}$ introducing a relative error in orientation of about 0.2 deg, but always constant. The earth angular velocity $\boldsymbol{\Omega}$ will be neglected with respect to the robot angular velocity, since its order of magnitude is too low, for the quantity to be measured. Finally the equations to be used are simply:

$$\dot{\boldsymbol{v}} = \boldsymbol{f} + \boldsymbol{g} - \rho \wedge \boldsymbol{v} .$$

4.2.3 Available Inertial Sensors

The elaboration of a method of analysis of inertial information is deeply function of the kind of sensors available, their precision, and their dynamic response. Inertial sensors are of two kinds: **linear accelerometers** and **gyrometers or gyroscopes** . The mechanical principle of these sensors is not to be discussed here (see for example [4.1] for an exhaustive review), but their performances, and the kind of error to be taken into account is to be analysed. This preliminary study is essential since all subsequent treatments will depend upon what is measured. Our discussion will be based on **low cost** sensors characteristics, for two practical reasons:

− their performances will be shown to be sufficient in a robotic environment,
− processing based on such sensors will *a fortiori* work better with improved sensors.

Linear Accelerometers. Linear accelerations induce inertial forces. These forces create displacements and these displacements can be measured. Available sensor relative precision (nonlinearity, hysteresis, reproducibility) is of about 0.5 % to 0.02 % with respect to the scale of measurement, while the resolution varies from 0.1 % down to 0.001 % for the best units. Robots are usually submitted to low accelerations of less than $1\,g \simeq 10$ m/s^2 even if high accelerations can occur for short periods of time (impacts, high-frequency transient vibrations), which are not to be measured. Then for accelerometers measuring about ± 2 g we get a resolution better than 0.02 m/s^2 and a precision better than 0.1 m/s^2 .

For this type of sensor accelerations are measured in a frequency range from 0 Hz to about 100 Hz. The true bandwidth of such sensor is $0\ldots 40$ Hz$\pm$ 1 %.

Such sensors measure the acceleration along a given axis, but are also sensitive to transversal accelerations by a relative factor of 2 %, and a geometrical calibration must be performed.

Gyrometers and Gyroscopes. Gyrometers measure the angular velocity. Such measurement is based on the measure of Coriolis forces for a mobile in linear motion, when an additional rotation occurs. The elimination of other accelerations (gravity, angular acceleration, linear acceleration) is due to appropriate combination of different elements.

Gyroscopes are made of a solid body with high angular velocity around an axis of symmetry, thus having an important kinetic momentum, which induces a high angular stability in presence of perturbations. Such a device is used to preserve a fixed inertial orientation, while the rest of the system is in motion. It must be noted that **both devices provide an information about angular velocity.** The first device output is the instantaneous angular velocity, while the second device output is the integration of the angular velocity during a period of time. Angular position information can be computed from gyrometers output by integration, while angular velocity information can be computed from gyroscopes output by derivation.

A large number of reasons have lead us to restrict our study to gyrometers: The theoretical results on gyrometers can be applied to gyroscopes since both devices provide the same physical information. Instantaneous motion is an important parameter on a robotic system, provided directly by gyrometers but not by gyroscopes. Numerical integration is more stable than numerical derivation. Absolute orientation can be partially estimated from the computation of the orientation of the gravity vector. In industrial systems gyrometers should be preferred, since they are less costly, and much more easy to implement.

Available gyrometers precision and dynamic response are very similar to those of low-cost accelerometers. They have a relative precision (linearity) of 0.1 % with a resolution better than 0.04 deg/s while they can measure angular velocities up to $100\ldots 300$ deg/s. They have a bandwidth of $0\ldots 50$ Hz $\pm$ 1 %. They are sensitive to transversal accelerations by a factor of less 0.02 (deg/s)/(m/s)^2, which has to be taken into account. Hysteresis is not measurable on such devices.

If angular velocity is integrated from a gyrometer output in order to provide an angular reference as in the case of a gyroscope, angular position could be subject to a drift of 3 deg/min., while gyroscopes could maintain angular position with a drift of only 10 deg/hour down to 0.1 deg/hour. Thus, gyroscopes have to be used if no other reference of angular position is present on the robot.

4.2.4 Comparison with the Human Vestibular System

The human vestibular system provides information about head angular velocity or acceleration, and head linear acceleration. A review of physiological data can be found elsewhere and only artificial and biological inertial sensors in term of performances and functions will be compared here.

The human inertial sensors resolution is similar to low-cost artificial sensors (0.05 m/s^2 for human linear accelerometers 0.1 deg/s at 0.1 Hz and 0.01 deg/s at 1 Hz for human angular sensors). The main difference is that the human angular sensor is an **angular accelerometer** not able to measure angular velocity at low frequencies.

In addition, the dynamic bandwidth of the human inertial sensors is about 0 . . . 2 Hz in term of accelerations, much lower than any artificial sensor. In fact, the biological layers act as very good mechanical filters for vibrations, impacts, etc. and the head is only submitted to vibrations in this frequency range. The situation is quite different on artificial rigid robots in which all vibrations are transmitted.

Several human vestibular important functions could also be implemented in a robotic system. They are:

- Stabilization of the gaze. The vestibulo–ocular-reflex, inducing compensatory eye movements, stabilizes the visual world on the retina. This reflex behaves as a multi-dimensional, adaptive, linear feed-back.
- Eye–head coordination. During head movements, the eyes have to be moved in synergy with the head. Vestibular information is used to correct the eye displacements, and such a control is, in fact, more efficient than if neck proprioception had been used.
- Control of posture. Postural control is a complex task, basically based on visuo–vestibular interactions. In the absence of vision, when standing, the vestibular evaluation of the vertical plays an important role in the control of posture.
- Self-motion perception, also elaborated by a cooperation between visual and vestibular cues.

At higher levels, the human brain, as most animal brains, uses inertial information in deep interaction with vision, in motion estimation, motor tasks, and for orientation.

Behavioral models for the vestibular functions are always described in terms of low frequency, sequential, elementary functions. The numerical implementation of such functions on an artificial system could be done using standard processing elements, the amount of computations being limited. The neuronal vestibular structures, contrary to visual structures for example, have a smaller degree of parallelism, the different levels of processing being sequential. Parallelism seems to be related, in this case, to a redundancy in terms of computation, yielding a noise reduction, and a protection in case of neuronal alteration.

In conclusion, artificial inertial sensors performances can be better than biological inertial sensors performances, which has two consequences: (1) An artificial inertial system can be used, in Neurophysiology, to simulate the vestibular system, providing a real robotic model of the vestibular system. (2) Inertial measurements can, on a robot, be used as the vestibular system in human, for the realization of complex tasks, in cooperation with artificial vision.

4.3 Auto-Calibration of Inertial Sensors

4.3.1 Presentation

This section describes how accelerometer and gyrometer signals can be related to an instantaneous estimation of self-motion and orientation. Computations are divided in two steps: **A calibration step**, detailed in this section: Using a realistic *model of the sensors*, a relation between the physical quantities and the sensor signal is proposed, while calibration parameters are identified. These parameters are determined through two experimental procedures, one for the *accelerometers intrinsic calibration*, the other for the *gyrometers intrinsic calibration*. During these procedures, particular orientations and motions are to be given to the set of sensors, while a recursive statistic algorithm improves the estimation of the calibration parameters and the related errors. **A real-time computation of self-motion and orientation**, developed in the next section.

4.3.2 Sensor Models

The set of sensors is assumed to form a rigid object. The frame of reference $(O, \boldsymbol{X}, \boldsymbol{Y}, \boldsymbol{Z})$ is attached to the sensors. Its movement is described through linear acceleration $\boldsymbol{\gamma}(O)$, and angular velocity $\boldsymbol{\omega}$. Angular velocity and linear acceleration, being 3D vectors, their measurement requires two sets of three sensors: the three angular velocity sensors are noted W_1, W_2 and W_3, while the sensory information is represented by $\boldsymbol{W} = (W_1, W_2, W_3)^{\mathrm{T}}$; the three accelerometer sensors are noted A_1, A_2 and A_3, while the sensory information is represented by $\boldsymbol{A} = (A_1, A_2, A_3)^{\mathrm{T}}$.

Accelerometers. Since it is not mechanically possible to put all three accelerometers at the same point, the acceleration on each accelerometer A_i is not $\gamma(O)$, but $\gamma(M_i)$, where M_i is the accelerometer location. We have

$$
\gamma(M_i) = \frac{d^2 \boldsymbol{OM_i}}{dt^2} = \frac{d\boldsymbol{V}(M_i)}{dt} = \frac{d(\boldsymbol{V}(0) + \boldsymbol{\omega} \wedge \boldsymbol{OM_i})}{dt} ,
$$
$$
\gamma(M_i) = \gamma(O) + \dot{\boldsymbol{\omega}} \wedge \boldsymbol{OM_i} + \boldsymbol{\omega} \wedge (\boldsymbol{\omega} \wedge \boldsymbol{OM_i}) ,
$$
$$
\dot{\boldsymbol{OM_i}} = \boldsymbol{V}(M_i) - \boldsymbol{V}(O) = \boldsymbol{\omega} \wedge \boldsymbol{OM_i} .
$$

Each sensor measures the projection of the accelerations along its eigenaxis. However, although the sensor is also sensitive to transversal accelerations, while the real axis is not necessarily known, the scalar measure is of the form

$$
\langle \gamma(M_i) + \boldsymbol{g}, \boldsymbol{n_i} \rangle
$$

where $\boldsymbol{n_i}$, with $\|\boldsymbol{n_i}\| = 1$, defines the direction of the projection which is roughly known, but has to be precisely determined. The measure $\langle \gamma(M_i) + \boldsymbol{g}, \boldsymbol{n_i} \rangle$ is related to $\gamma(O)$ and $\boldsymbol{\omega}$ by

$$
\langle \gamma(O) + \boldsymbol{g}, \boldsymbol{n_i} \rangle + \langle \dot{\boldsymbol{\omega}}, \boldsymbol{OM_i} \wedge \boldsymbol{n_i} \rangle + \langle \boldsymbol{\omega} \wedge \boldsymbol{n_i}, \boldsymbol{\omega} \wedge \boldsymbol{OM_i} \rangle .
$$

Since accelerometers are mounted on a trihedral support, on which the direction of the projection $\boldsymbol{n_i}$ is close to the direction of $\boldsymbol{OM_i}$, the previous equation can be simplified.

Quantitatively, actual sizes of the sensor allow a design with $\|\boldsymbol{OM_i}\| \simeq$ 5 cm with $(\widehat{\boldsymbol{OM_i}, \boldsymbol{n_i}}) < 2$ deg while $(\boldsymbol{n_1}, \boldsymbol{n_2}, \boldsymbol{n_3})$ are orthogonal with a precision better than 2 deg. For angular velocities $\|\boldsymbol{\omega}\| \leq 50$ deg/s at frequencies up to 1 Hz the tangential acceleration $(\dot{\boldsymbol{\omega}} \wedge \boldsymbol{OM_i})$ and centrifugal acceleration $[\boldsymbol{\omega} \wedge (\boldsymbol{\omega} \wedge \boldsymbol{OM_i})]$ projections are bounded by

$$
| \langle \dot{\boldsymbol{\omega}} \wedge \boldsymbol{OM_i}, \boldsymbol{n_i} \rangle | \leq 0.015 \text{ m/s}^2 ,
$$
$$
| \langle \boldsymbol{\omega} \wedge (\boldsymbol{\omega} \wedge \boldsymbol{OM_i}), \boldsymbol{n_i} \rangle | \leq 0.15 \text{ m/s}^2 .
$$

The first quantity is negligible (below the sensor resolution), while the second has to be taken into account. However, since this last quantity is small, and since $\boldsymbol{n_i}$ is close to the direction of $\boldsymbol{OM_i}$ the normal acceleration factor can be simplified. We are going to assume $\boldsymbol{OM_i} \simeq \|\boldsymbol{OM_i}\| \cdot \boldsymbol{n_i}$ and $\|\boldsymbol{OM_1}\| \simeq \|\boldsymbol{OM_2}\| \simeq \|\boldsymbol{OM_3}\| \simeq \delta$, as designed mechanically. We have finally for $\langle \gamma(M_i) + \boldsymbol{g}, \boldsymbol{n_i} \rangle$:

$$
\langle \gamma(O) + \boldsymbol{g}, \boldsymbol{n_i} \rangle + \delta \cdot (\|\boldsymbol{\omega}\|^2 - \langle \boldsymbol{\omega}, \boldsymbol{n_i} \rangle^2) . \tag{4.1}
$$

The digital value, after acquisition by the system, is related to this quantity, by a linear relation of the form:

$$
A_i = O_i + (1 + K_i) \cdot \langle \gamma(M_i) + \boldsymbol{g}, \boldsymbol{n_i} \rangle
$$

where the offset O_i and the gain $(1 + K_i)$ describe the signal transformation through the electronic acquisition system. Their adjustments will provide a reference for the zero, and a scaling factor.

Three arguments are in favour of the use of such a linear relation in order to approximate the electronic response:

- The errors are not related to nonlinearities, but to non repetitive, non deterministic factors (noise), and the measurement on accelerometers appears to have a better linearity than its resolution.
- Electronic performances are much better than the sensor precision, which is about 0.1 % $\Longleftrightarrow$ 60 dB, while the acquisition is done with 12 ebs $\Longleftrightarrow$ 72 dB, with equivalent performances for electronic interfaces, in terms of noise and linearity.
- The main unknowns are the scale and the zero, known with an accuracy of less than 1 %.

Gyrometers. The analysis of gyrometer responses is very similar to the analysis of accelerometers, but much simpler since, on a rigid object, the angular velocity is not dependent upon the point where the measure is taken, and

$$\boldsymbol{\omega}(M_i') = \boldsymbol{\omega}(O) = \boldsymbol{\omega} \ .$$

The only complication comes from the fact that velocity rate sensors are also sensitive to transversal linear accelerations, the digital value being related to both angular velocity and linear accelerations by an equation of the form

$$W_i = O_i' + (1 + K_i') \cdot \langle \boldsymbol{\omega}, \boldsymbol{n_i}' \rangle + L_i' \cdot \langle \boldsymbol{\gamma}(M_i') + \boldsymbol{g}, \boldsymbol{m_i}' \rangle$$

where the offset O_i' and the gain $(1 + K_i')$ describe the signal transformation through the electronic acquisition system, while $\boldsymbol{n_i}'$ is the axis of the sensor. L_i' and $\boldsymbol{m_i}'$ are the gain and the direction of the linear acceleration component acting on the sensor. From the mechanical design of the sensor we can expect $\boldsymbol{m_i}'$ to be orthogonal to $\boldsymbol{n_i}'$.

Equation (4.1) couples $\boldsymbol{\omega}$ and $\boldsymbol{\gamma}$, and is nonlinear. Due to this equation, the calibration of a sample might yield to the resolution of a quadratic equation in $\boldsymbol{\omega}$. However, L_i' being expected to be small ($\simeq$ 1 %) the acceleration $\boldsymbol{\gamma}(M_i') + \boldsymbol{g}$ can be approximated by $\boldsymbol{\gamma}(O) + \boldsymbol{g}$ in the previous equation, because the residual normal acceleration, $\delta \cdot (\|\boldsymbol{\omega}\|^2 - \langle \boldsymbol{\omega}, \boldsymbol{n_i} \rangle^2)$, multiplied by L_i' is expected to be negligible($< 10^{-4}$).

Therefore when no motion is present, W_i is simply given by the following equation:

$$W_i = O_i' + L_i' \cdot \langle \boldsymbol{g}, \boldsymbol{m_i}' \rangle = O_i' + \sum_{k=1}^{3} \alpha_i'^k \cdot \langle \boldsymbol{g}, \boldsymbol{n_k} \rangle \tag{4.2}$$

where $\alpha_i'^k = L_i' \cdot m_i'^k$, and $m_i'^k$ are the contravariant coordinates of $\boldsymbol{m_i}'$ in the $\boldsymbol{n_i}$ frame of reference.

When the system is in motion, let us note W_i' *the gyrometer digital value when offset and acceleration contaminations are eliminated.* We have

$$W_i' = W_i - O_i' + \sum_{k=1}^{3} \alpha_i'^{k} \cdot \langle g, n_k \rangle = \sum_{k=1}^{3} \beta_i'^{k} \cdot \langle \omega_k, n_k \rangle \qquad (4.3)$$

using similar transformations.

Remarks: As for accelerometers the nature of the acquisition system allows the use of a linear model of measurement for the sensor. We do not use the constraint $\langle n_j', m_i' \rangle = 0$ because it is based on an assumption concerning the sensor behaviour which might not be entirely true and this additional constraint would increase the complexity of the calibration equations, since it is quadratic in terms of the calibration parameters.

Dynamic Responses. In a frequency range of 0 to 50 Hz, the frequency response in term of gain and phase is nearly perfect, and the resulting is less than other sources of errors. It is then possible, not to take the dynamic responses into account assuming a gain of one and a phase lag or lead of zero.

In fact, for higher frequencies of vibrations, mechanical adequate structures acting as passive filters would have better responses, in term of stabilization, than active mechanisms, based on the use of these sensors.

Calibration Parameters. The calibration parameters to be estimated are:
For the linear accelerometers: the gains and offsets: K_i and O_i ; the inner products: $\langle n_j, n_i \rangle$, $i \neq j$.
For the gyrometers: the offsets and acceleration contamination gains O_i' and $\alpha_i'^{k}$; the gain matrix $\beta_k'^{i}$.

4.3.3 Accelerometers Intrinsic Calibration

The gravity field is a very good reference for accelerometers calibration. This acceleration vector is constant in direction and amplitude, and does not depend upon the accelerometer location in space. These properties will be used to calibrate, without any special hardware, the accelerometers by simply orienting them in particular directions with respect to gravity.

4.3.4 Static Evaluation of Calibration Parameters

When no motion is present, the measurement relation is simply related to the gravity vector g:

$$A_i = O_i + (1 + K_i) \cdot \langle g, n_i \rangle$$

and it is obvious that A_i is maximum when g is aligned with n_i and has the same direction, while A_i is minimum when g is aligned with n_i and in the opposite direction. Using this property, let us note A_i^{j+} the acceleration on the axis i when A_j is maximal, and A_i^{j-} the acceleration on the axis i when A_j is minimal. We then get

$$A_i^{j+} = O_i + (1 + K_i) \cdot ||g|| \cdot \langle n_j, n_i \rangle \,,$$
$$A_i^{j-} = O_i - (1 + K_i) \cdot ||g|| \cdot \langle n_j, n_i \rangle \,.$$

Then, calibration parameters can be calculated by the following formulas, easily derived:

$$O_i = \frac{A_i^{j+} + A_i^{j-}}{2} \qquad \text{for} \qquad j = 1, 2, 3 \,, \tag{4.4}$$

$$1 + K_i = \frac{A_i^{i+} - A_i^{i-}}{2 \cdot ||g||} \,, \tag{4.5}$$

$$\langle n_j, n_i \rangle = \langle n_i, n_j \rangle = \frac{A_i^{j+} - A_i^{j-}}{A_i^{i+} - A_i^{i-}} = \frac{A_j^{i+} - A_j^{i-}}{A_j^{j+} - A_j^{j-}} \,. \tag{4.6}$$

In order to simplify the notations we will now note $\mu_{ij} = \langle n_j, n_i \rangle$. These terms are in fact the coordinates of the metric tensor associated with the non-orthogonal frame of reference (n_1, n_2, n_3).

The previous equations allow the calculation of the acceleration in this non-orthogonal frame of reference, in which:

$$\langle \gamma(O) + g, n_i \rangle = \frac{A_i - O_i}{1 + K_i} - \delta \cdot (||\omega||^2 - \langle \omega, n_i \rangle^2) \,. \tag{4.7}$$

It will however be useful to get the vector coordinates in an orthogonal frame of reference. Computing scalar and dot products will be costless, while the relation between two orthogonal frames is easily derived. The matrix transformation is only function of μ_{ij} and will be given in the next section.

At last, the parameter δ can be measured directly on the mechanical hardware, with a relative precision of 1 %. Since it is used only in the correction factor of (4.7), this direct estimation is sufficient.

4.3.5 Experimental Procedure for Accelerometers Calibration

The estimation of the calibration parameters requires only to be able to orient the accelerometers in any direction, and to maintain the system without any movement, during the static measurements.

The estimation of A_i^{j+} (or A_i^{j-}) can be performed as follows. Let $p = (p_1, p_2)$ be the parameters of any representation of the orientation of the axis of the accelerometer i. One should remember that *the orientation of an axis in 3D space is defined by two parameters*. The static measurement A_i on accelerometer i, is only a function of p: $A_i(p)$. The estimation of A_i^{i+} is simply a two degree of freedom non-constrained maximization problem of the form

$$A_i^{i+} = \max_{p = (p_1, p_2)} \{A_i(p)\}$$

while the estimation of A_i^{j-} is given by a similar problem:

$$A_i^{i-} = \max_{p=(p_1,p_2)} \{-A_i(p)\} \ .$$

This maximum can be obtained using any nonlinear optimization method. Such a method will converge since $A_i(p)$ is a priori a convex function, with only one maximum, if $A_i(p)$ is in the neighborhood of A_i^{i+}. In practice, it is very easy to manually orient the accelerometer more or less in the vertical direction, and then apply an optimization method in order to get a precise value of the maximum. Every optimization method (gradient, Newton, Powell, etc.) has the following common structure:

(1) Start from an initial reasonable p (for which the accelerometer is more or less in the vertical direction).
(2) Measure $A_i(p)$ for a given p.
(3) Using a set of measurements of $A_i(p)$, find a better p.
(4) Repeat step 2–3 until $A_i(p)$ cannot be increased.

Implementation of such a method is standard, and not discussed here.

The following experimental procedure, based on the previous discussion, is rather simple and can be totally automated, using for example a robotic arm.

---

(1) For each axis orient the accelerometers to get the highest signal on the sensor output, and measure, on each sensor, the three outputs which correspond to A_i^{j+}.
(2) For each axis orient the accelerometers to get the lowest signal on the sensor output, and measure, on each sensor, the three outputs which correspond to A_i^{j-}.
(3) Compute O_i, K_i, and $\mu_{ij} = \langle n_i, n_j \rangle$ using (4.4), (4.5), (4.6).
(4) Compute M_j^i from μ_{ij}.
(5) Measure the precision of the calibration by comparing (4.4) together and (4.6) together, each set of equations should be verified with a precision close to sensor resolution (10^{-3}).
(6) Measure δ, on the mechanical support.

Calibration procedure for accelerometers

4.3.6 Gyrometers Intrinsic Calibration

Parameters of (4.2) and (4.3) are to be computed in order to be able to compute angular velocity from gyrometers measurements. This will be done during two separate procedures, as detailed now.

Gyrometers Static Calibration. When no motion is present, the accelerations are only related to g. In addition, the angular velocity should be zero, and the value of W_i' in (4.2) should also be zero. This provides for each measurement of W_i and A_k, a linear equation in term of the calibration parameters O_i' and $\alpha_i'^k$.

As soon as four independent measurements of (W_i, A_1, A_2, A_3) are known, this set of equations can be solved. Such independent measurements can be obtained during the accelerometers calibration procedure.

In fact, a recursive mean-square estimation will be chosen, since it is a good method to evaluate these parameters, even if there is some noise. The precision of the estimation (in terms of covariance) is also given by such an algorithm.

The complete calibration procedure is simply:

---

(1) During the accelerometers calibration procedure, each time a measurement of (A_1, A_2, A_3) is taken, while the system is not in motion, measure also W_i.
(2) Using this measurement, perform a recursive mean-square adjustment of O_i' and $\alpha_i'^k$.

Calibration procedure of the gyrometers offsets and acceleration contaminations

Gyrometers Dynamic Calibration. We will now assume W_i' to be computed from W_i and A_k for each sample. The accelerometers being calibrated, g can be computed each time the system is not in motion. We are now dealing with the computation of the calibration parameters $\beta_k'^i$ of (4.3).

This equation is linear in term of $\beta_k''^i$, and only ω is to be determined. Since we want to develop a general method, not only related to a specific equipment, ω will be calculated from already calibrated inertial information, mainly g. We will see that particular rotations have to be used, and we will show how to generate them.

4.3.7 Principle of the Dynamic Calibration

During a rotation, the gravity vector's time derivative is simply related to angular velocity by

$$\frac{\mathrm{d}g}{\mathrm{d}t} = \omega \wedge g$$

since $||g|| = 1$ $g \simeq 9.81$ m/s^2 and is constant. If the direction of ω remains constant and orthogonal to g (rotation performed in a vertical plane), the angular velocity is related to the gravity and the gravity time derivative by

$$\omega = \frac{1}{||g||^2} \cdot g \wedge \frac{\mathrm{d}g}{\mathrm{d}t} \qquad \text{if} \qquad \omega \perp g . \tag{4.8}$$

It might then be possible, *for a rotation around a fixed axis performed in a vertical plane*, to calculate the angular velocity from the measurement of g.

4.3.8 Computing the Rotation from g

In order to have an efficient estimation, we are not going to use (4.8) directly, but integrate several measurements during a time interval $[0, T]$. Since each rotation is performed around a fixed axis, represented by u, we have after a rotation of direction u, of angle Θ, during a time T:

$$\omega(t) = \dot{\Theta}(t) \cdot u ,$$

$$\Theta = \int_0^T \dot{\Theta}(t) \mathrm{d}t$$

and we obtain:

$$\int_0^T W_i'(t) \mathrm{d}t = \Theta \cdot \sum_{i=1}^3 \beta_k'^i \cdot \langle u, n_k \rangle . \tag{4.9}$$

If $g(0)$ and $g(T)$ are the measurements of g at the beginning and the end of the rotation, we have, since g remained orthogonal to the axis of rotation:

$$\cos(\Theta) = \frac{1}{||g||^2} \cdot \langle g(0), g(T) \rangle ,$$

$$u = \frac{g(0) \wedge g(T)}{||g(0) \wedge g(T)||} \tag{4.10}$$

yielding:

$$\Theta \cdot \langle u, n_k \rangle = \arccos \left(\frac{\langle g(0), g(T) \rangle}{||g||^2} \right) \cdot \frac{\langle g(0) \wedge g(T), n_k \rangle}{||g(0) \wedge g(T)||} . \tag{4.11}$$

Equation (4.9) is in fact an integral form of (4.3), when rotation is performed around a fixed axis. It has three advantages:

- This equation is still linear in terms of $\beta_k'^i$ and can also be solved using a recursive mean-square estimation.
- Using (4.11) it is then possible to compute $\Theta \cdot \langle u, n_k \rangle$ from $g(0)$, $g(T)$.
- Averaging several measurements, the estimation will be more reliable.

It is important to note that g is measured at the beginning and at the end of the movement, *when the system is not in motion*. The coupling between ω and g, as given in (4.1) is then avoided. This is mandatory, since ω is not yet known.

These equations are verified only if the rotation is performed in a vertical plane and we are now showing how to orient the axis of rotation in order to match this requirement.

4.3.9 Performing a Rotation in the Vertical Plane

We assume that we are given a mechanical axis of rotation, we can orient, in any direction, since we can perform with it a given rotation, around a given axis, and of a given angle. But this mechanical system is not necessarily calibrated and *we will not rely on external measurements*. In addition, inertial sensors might be more precise than usual mechanical measurements devices, and even if calibrated the obtained measurements might not have enough precision.

Let us make a rotation with *a fixed angle* Θ_0. Using (4.10), the angle of rotation computed from g is

$$\Theta = \arccos\left(\frac{\langle g(0), g(T)\rangle}{||g||^2}\right) .$$

If the rotation is not performed in the vertical plane, there is also a rotation around g, of some angle, and we get: $\Theta \leq \Theta_0$. We have $\Theta = \Theta_0$, if and only if, the rotation is really in the vertical plane. In other words, Θ *is maximum if and only if the rotation is performed in the vertical plane.*

Let $p = (p_1)$ be the parameters of any representation of the orientation of the axis of rotation with respect to the horizontal plane. One should remember that *the orientation of an axis in 3D space with respect to a plane is defined by one parameter*. The adjustment of the axis of rotation is simply a one degree of freedom non-constrained maximization problem of the form

$$\Theta_0 = \max_{p=(p_1)}\{\Theta(p)\}$$

This maximum can be obtained using any nonlinear optimization method. Such a method will converge since this function is, in a neighbourhood of the maximum, a convex function, with only one maximum. If the axis is roughly vertical, as it is very easy to obtain in practice, this method will converge.

4.3.10 Experimental Procedure for Calibration

The following experimental procedure, based on the previous discussion, summarizes the different steps for the gyrometers dynamic calibration.

---

(1) Choose an orientation of the axis of rotation with respect to the gyrometers.
(2) Perform a rotation.
(3) Modify the orientation of the axis of rotation in order to maximize Θ.
(4) Repeat steps 2 to 3 until Θ is maximum.
(5) Perform a rotation.
(6) Compute $\Theta \cdot \langle u, n_1\rangle$ using (4.11).
(7) Use (4.9) to recursively estimate $\beta_k'^i$ and the related covariances.
(8) Repeat steps 5 to 7 until the estimation is correct.
(9) Repeat steps 1 to 8 for three independent axis.

Calibration procedure for gyrometers gains

4.3.11 Experimental Results for Calibration

The measurements for two calibration sessions are given here. Calibrations
have been done on the same system, on two different days. Parameters are
estimated with their precision, given either as an error (ϵ), or a standard
deviation (σ).

Accelerometers gains G and offsets O are given. The accelerometers geometry is measured through the μ matrix, ideally symmetric. In our mechanical
construction, the orthogonality of the sensors is rather high, and the scalar
products in μ are negligible.

Gyrometers offsets O'' and relations to accelerations α'' are given. It
should be noted that α'' values and standard-deviations are, for clarity, multiplied by 100. Gyrometers gains are omitted here but yield similar values.

$$O = \begin{pmatrix} -286\ (\epsilon = 6) & -633\ (\epsilon = 5) & -462\ (\epsilon = 5) \end{pmatrix}$$

$$G = \begin{pmatrix} 1262 & 1162 & 1253 \end{pmatrix}$$

$$\mu = \begin{pmatrix} 1 & -0.0046 & 0.0047 \\ -0.0027 & 1 & -0.0010 \\ 0.0123 & -0.0013 & 1 \end{pmatrix}$$

$$O' = \begin{pmatrix} -1344\ (\sigma = 26) & 734\ (\sigma = 15) & -462\ (\sigma = 9) \end{pmatrix}$$

$$100 * \alpha' = \begin{pmatrix} 4.3\ (\sigma = 0.6) & -1.3\ (\sigma = 0.3) & 0.9\ (\sigma = 0.1) \\ 2.5\ (\sigma = 0.1) & 3.7\ (\sigma = 0.1) & -1.2\ (\sigma = 0.1) \\ 1.6\ (\sigma = 0.1) & -0.9\ (\sigma = 0.1) & 4.2\ (\sigma = 0.3) \end{pmatrix}$$

Results of a first calibration session

$$O = \begin{pmatrix} -286\ (\epsilon = 2) & -616\ (\epsilon = 4) & -490\ (\epsilon = 5) \end{pmatrix}$$

$$G = \begin{pmatrix} 1263 & 1166 & 1254 \end{pmatrix}$$

$$\mu = \begin{pmatrix} 1 & -0.0057 & 0.0078 \\ -0.0070 & 1 & -0.0014 \\ 0.0056 & -0.0009 & 1 \end{pmatrix}$$

$$O' = \begin{pmatrix} -1274\ (\sigma = 24) & 765\ (\sigma = 13) & -412\ (\sigma = 7) \end{pmatrix}$$

$$100 * \alpha' = \begin{pmatrix} 4.2\ (\sigma = 0.4) & -1.6\ (\sigma = 0.2) & 1.1\ (\sigma = 0.2) \\ 2.1\ (\sigma = 0.1) & 3.3\ (\sigma = 0.2) & -1.5\ (\sigma = 0.1) \\ 2.0\ (\sigma = 0.3) & -0.9\ (\sigma = 0.1) & 4.1\ (\sigma = 0.2) \end{pmatrix}$$

Results of a second calibration session

4.4 Separation Between Gravity and Linear Acceleration

Gravity and linear acceleration are two quantities which can not be differentiated with any physical experiment. However, in a robotic environment, special assumptions can be used. Based on these assumptions, additional constraints can be used to separate gravity and linear acceleration. We first motivate two general assumptions on these quantities, and then propose two related methods of separation.

4.4.1 Using Special Assumptions on the Environment

The accelerations measured by the linear accelerometers are the sum of the object acceleration and the gravity. There is no possibility, *a priori*, to separate linear acceleration and gravity, but several additional hypotheses can be used, depending of the context. For example, for a mobile robot running on an floor, one can assume that acceleration is in the plane of the wheels, except if in a lift, or during a free fall etc. In addition, the direction of the movement is given by the wheels directions, assuming skid does not occur.

However there are two physically realistic hypotheses, not dependent upon the object specific movement:

(1) **g is a constant uniform 3D vector field**. Then, as already formulated:

$$\frac{\mathrm{d}g}{\mathrm{d}t} = \omega \wedge g .$$

(2) **It is not physically possible to keep acceleration constant**. Then, for a relatively large window of time, and in an absolute frame of reference, the object acceleration belongs to the higher part of the acceleration frequency spectrum, while gravity is related, in theory, to the lower part of the acceleration frequency spectrum. This means that the DC component of $\gamma(O)$ is close to 0 if the time-window is large enough, and we have

$$g = [\gamma(O) + g] * \int_{-\infty}^{+\infty} I \cdot \delta_0(f) \cdot \mathrm{e}^{-2\pi \mathrm{j} f t} \mathrm{d} f$$

where $*$ denotes the convolution operator, I the identity matrix, and vector coordinates are expressed in an absolute frame of reference.

We now describe two methods only based on these hypotheses, and which should be rather general, and not only restricted to, for instance, vehicle movements.

4.4.2 Method 1: Estimating $\gamma(O)$ in an Absolute Frame of Reference

Using the estimation of the angular rotation through $\lambda(t)$ we can also express $a = \gamma(O) + g$ with respect to the initial orientation. More precisely, a vector attached to the initial frame of reference is related to a vector attached to the actual frame of reference by:

$$a_{\text{absolute}} = R(t)^{\text{T}} \cdot a_{\text{inertial}} = \bar{\lambda}(t) \times a_{\text{inertial}} \times \lambda(t)$$

since $R(t)$ is the matrix of the rotation from the initial frame to the actual frame, both frames being orthogonal. One should, of course, remember that for an orthogonal matrix R: $R^{\text{T}} = R^{-1}$, while for a unitary quaternion λ: $\bar{\lambda} = \lambda^{-1}$ [4.2–4].

In this absolute frame of reference we have

$$a_{\text{absolute}} = R(t)^{\text{T}} \cdot \gamma(O) + g(0)$$

where $g(0)$ is the fixed value of the gravity field. If $R(t)^{\text{T}}$ had been estimated with no errors, one could write

$$\gamma(O)_{\text{inertial}} = a_{\text{inertial}} - R(t) \cdot g(0) \ .$$

This is however not the case, and since the estimation of $R(t)$ is subject to a drift, a filter has to be used.

The second hypothesis provides also an estimation of the object acceleration noted $\tilde{\gamma}(O)$. It is just necessary to filter $a = \gamma(O) + g$ measured by the accelerometers, in order to eliminate the "zero-frequency" component assumed to be g, if the vectors are taken in an absolute frame of reference. We have

$$
\begin{aligned}
\tilde{\gamma}(O) &= \lambda(t) \times \left\{ F(t) * \left[\bar{\lambda}(t) \times a \times \lambda(t) \right] \right\} \times \bar{\lambda}(t) \ , \\
\tilde{g} &= a - \tilde{\gamma}(O)
\end{aligned}
\tag{4.12}
$$

while

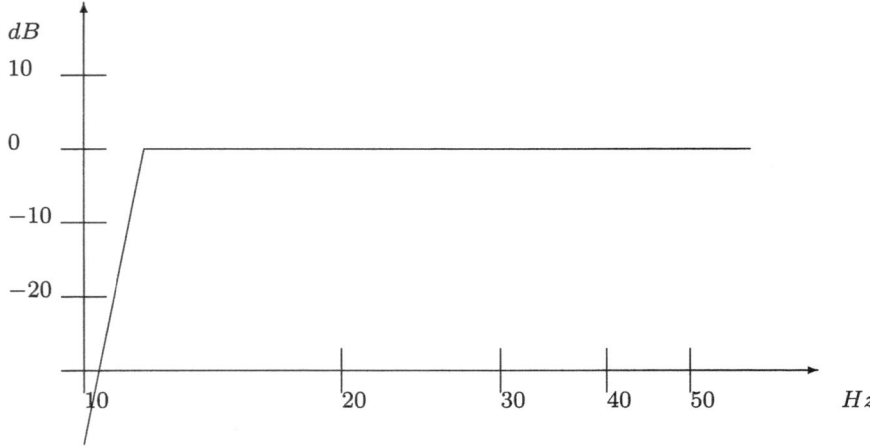

Fig. 4.2. Profile of $F(f)$ gain as a function of frequency (see text for details)

$$F(t) \simeq \int_{-\infty}^{+\infty} I \cdot [1 - \delta_0(f)] \cdot e^{-2\pi j f t} df$$

is the approximation of the impulse response of "frequency-zero" rejection filter. The shape of such a filter is shown in Fig. 4.2.

The error on $\lambda(t)$ is mainly a slow drift, that is a low frequency component of $\lambda(t)$, and is filtered by $F(t)$ in (4.12). Then, contrarily to what happens when integrating ω, this estimation of g for which very low frequencies are filtered out is not subject to drift.

4.4.3 Method 2: Estimating $\gamma(O)$ Using the Jerk

The jerk j is the time derivative of the accelerations and we have

$$j = \dot{a} = \dot{\gamma}(O) + \dot{g} = \dot{\gamma}(O) + \omega \wedge g = \dot{\gamma}(O) + \omega \wedge [a - \gamma(O)]$$

which provides a differential equation in term of $\gamma(O)$:

$$\dot{\gamma}(O) = \omega \wedge \gamma(O) + \dot{a} - \omega \wedge a \ .$$

Such an equation can be solved using different methods, either used in the internal model of a Kalman filter, or in a simple feedback as given on Fig. 4.3. This feedback is a nonlinear filter and will be noted:

$$\gamma(O)(t) = G\left(a(t), \omega(t)\right) \ .$$

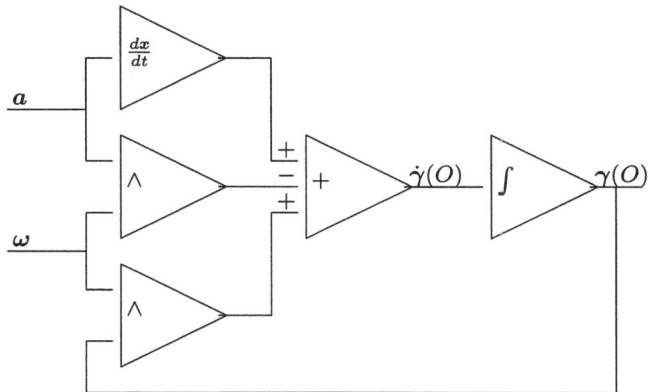

Fig. 4.3. Feedback mechanism for the estimation of $\gamma(O)$

With this equation one can separate the two components of the specific forces, and we obtain g and $\dot{v}$ from two differential equations, with initial integral conditions:

$$\begin{aligned} \ddot{v} &= \dot{a} + \omega \wedge (\dot{v} - a) & \int_{-\infty}^{0} \dot{v}(t)dt &= 0 \ , \\ \dot{g} &= \omega \wedge g & \int_{-\infty}^{0} g(t)dt &= \int_{-\infty}^{0} a(t)dt \ . \end{aligned}$$

We thus have a long term estimate of the vertical orientation from this sensor.

We have a short term estimate of the horizontal and vertical orientations from this sensor, thus a short term estimate of the vertical.

Both sets of sensors have, in fact, a precision similar to the vestibular system of the mammalian, which act as a inertial system. A comparison is given in [4.5].

In addition, there is natural decomposition of self-motion in terms of vertical and horizontal orientation, when using inertial cues alone [4.6]. Vertical orientation is estimated without any drift because of the use accelerometers measuring g, while errors in horizontal orientation increase with time.

4.5 Integration of Angular Position

We now deal with the problem of integrating ω in order to obtain an estimation of the angular position.

The attitude of a rigid body is given by a 3D rotation, the set of 3D rotations being isomorphic to the group SO_3, while rigid body position is given by the Cartesian product of a rotation and a translation, the set of rigid displacements being isomorphic to the group SE_3 [4.7]. This last group, as a manifold, has a tangent space se_3, isomorphic with the set of screws, that is a combination of linear and angular velocities. A detailed introduction is in [4.8]. The fact we are "in the discrete case" means we are not going to work with velocities (se_3), but positions (SE_3). This is because visual sensor data are sampled at a rate which is too slow to "linearize" the equations.

It is not obvious to parameterize the 3D rotation which has 3 degrees of freedom, and to relate this representation to the matrix R. Several solutions exist (Euler angles, orthogonal matrix with positive determinant (positive isometry), unary quaternions, antisymmetric matrix exponential and Cayley formula, Rodriguez formula and compact representation, etc.) [4.2, 9]. Whatever the representation is, the main problem is that the 3D rotation manifold (isomorphic to the group $SO3$) has not a trivial topology, and need more than one map in its atlas (no global parametrization). However, if we restrain to "non degenerated" rotations this problem disappears (degenerated rotations are rotations with an angle equal to $\pm\pi$, thus equivalent to a symmetry with respect to a line, this subset being isomorphism to half of the Gauss sphere S^2). This is reasonable if we consider small rotations, as when considering the motion between two consecutive frames. The subspace of non-degenerated 3D rotations is isomorphic to $\mathcal{R}^3$. In any case, the desirable properties are: minimal representation (thus of dimension 3), rational representation (to simplify algebraic developments) with close form (preferably simple) formulas, one-to-one mapping between the representation and the related matrix, simple approximations when considering small rotations.

4.5.1 A Rational Representation of 3D Rotations

One compromise, which corresponds the previous requirements and is used in our implementation, is:.

Proposition 4.1. The matrix[1]

$$R(v) = I + \left(\frac{\tilde{v} + \frac{1}{2}\tilde{v}^2}{1 + \frac{1}{4}v^{\mathrm{T}} \cdot v} \right)$$

is a non-degenerated 3D rotation matrix of axis direction $u = v/\|v\|$ and of angle $\theta = 2\arctan(\|v\|/2)$.

On the reverse, considering a 3D rotation matrix corresponding to a non-degenerated rotation of angle $\theta \in]-\pi, \pi[$ and of axis direction u, there is a unique vector $v = 2\tan(\theta/2) \cdot u/\|u\|$ for which the rotation matrix is given by the previous relation. We also have the inverse relation $v = 2\tan[\arcsin(\|w\|)/2] \cdot w/\|w\|$, with $\tilde{w} = (R - R^{\mathrm{T}})/2$.

Moreover:

(1) The first and second order expansions of R are straightforward with this representation since we have: $R = I + \tilde{v} + o(\theta^2) = I + \tilde{v} + \frac{1}{2}\tilde{v}^2 + o(\theta^3)$, and $v = w + o(\theta^3) = \theta \, u + o(\theta^3)$.

(2) The Jacobian of the matrix with respect to the representation vector is, for small rotations:

$$\frac{\partial R_{ij}}{\partial v^k} = \epsilon_{ijk} + o(\theta^2) \, ,$$

where ϵ_{ijk} is the Eddington symbol[2].

(3) This representation includes the null rotation $R = I$. A degenerated rotation of axis direction u, $\|u\| = 1$ and of angle $\pm\pi$ is similarly parameterized by $R = I + 2\tilde{u}^2 = \lim_{\|v\| \to \infty, u = v/\|v\|} R(v)$.

Proof. Slightly heavy, but straightforward. We just give the method.

Compute $R \cdot R^{\mathrm{T}}$ to verify we have an isometry. Compute $R \cdot v$ to verify that v is invariant thus aligned with the rotation axis. Verify no other vector direction could be invariant, thus showing this is not a symmetry but a true rotation (positive isometry). Compute $x^{\mathrm{T}} \cdot (R \cdot x) = \cos(\theta)$ for $x \perp v$ to relate

---

[1] We write

$$\tilde{u} \cdot x = \begin{pmatrix} 0 & -u_z & u_y \\ u_z & 0 & -u_x \\ -u_y & u_x & 0 \end{pmatrix} \cdot x = u \wedge x \, .$$

The identity matrix is written I.

[2] The Eddington (or Levi-Civita) symbol is defined by the following relations

$$\epsilon_{ijk} = \begin{cases} 0 & \text{if} \quad i = j \quad \text{or} \quad i = k \quad \text{or} \quad j = k \\ 1 & \text{if} \quad i < j < k \quad \text{or} \quad k < i < j \quad \text{or} \quad j < k < i \\ -1 & \text{if} \quad i < k < j \quad \text{or} \quad k < j < i \quad \text{or} \quad j < i < k \end{cases} .$$

It is an antisymmetric tensor of type 3 and obviously $w = u \wedge v$ can be computed as $w_k = \sum_{i,j} \epsilon_{ijk} u^i v^j$, while $\tilde{u}_{jk} = \sum_i \epsilon_{ijk} u^i$.

the rotation angle to the magnitude of v with a few trigonometry, considering $\phi = \theta/2$. In particular $c = 1/[1 + (v^{\mathrm{T}} \cdot v)/4] = \cos(\theta/2)^2$.

On the reverse, consider the Rodriguez formula for a 3D rotation $R = I + \sin(\theta)\tilde{u} + [1 - \cos(\theta)]\tilde{u}^2$ [4.2], and expand trigonometric relations in function of $\theta/2$ to identify with the present formulas.

First order and second order expansions are also obtained considering expansions in $\epsilon = \theta/2$. The formula to obtain the variation of R with respect to v is obtained from the first order expansion and is trivial.

Degenerated rotations are symmetries of axis u, as it can be easily verified for a vector parallel to u and a vector orthogonal to u on the related formula. This last formula could also have been derived from the Rodriguez formula.

We, in fact, have verified all these results on a symbolic calculator. ☐

This rational, minimal, one-to-one, representation is very attractive because it yields simple formulas especially for small rotations and computation of covariances. In fact, it is a variation of the antisymmetric exponential representation $R = \exp(\tilde{r}), r = \theta u$, but with a nonlinear transform of the vector magnitude which dramatically simplifies the formulas.[3] It also provides an isomorphism between non-degenerated rotations and $\mathcal{R}^3$.

4.5.2 Relation Between ω and the Angular Position

Let us now discuss the relation between continuous velocities and accelerations, and small displacements. This corresponds to considering a second order approximation of $R(v)_t$ at time t, when t is infinitesimal. We consider $R_0 = I$. Let ω_t be the angular velocity at time t and $\omega = \omega_0$ and $\dot{\omega} = \dot{\omega}_0$. Let $R_t^{\bullet}$ be the rotation matrix that represents the discrete rotation between time 0 and t. It is known to satisfy the following differential equation [4.10]

$$\frac{\partial R_t^{\bullet}}{\partial t} = \omega_t \cdot R_t^{\bullet} .$$

Taking into account the fact that we are interested in infinitesimally small values of t, we look for a solution of this equation by using a second order Taylor series expansion of $R_t^{\bullet}$ and obtain

$$R_t^{\bullet} = I + t[\tilde{\omega}] + \frac{t^2}{2}\left[\tilde{\omega}^2 + \tilde{\dot{\omega}}\right] + o(t^3) .$$

On the other hand, considering $v_t = v + \delta v$ be value of v at time t, with $\|v\| = o(t)$ (small rotation) we have

$$R(v_t) = I + \tilde{v} + t\tilde{\delta v} + \frac{1}{2}\tilde{v}^2 + o(t^3) .$$

If $R_t^{\bullet}$ corresponds to $R(v_t$ up to the second order for small t, this leads to $v = t\,\omega$ and $\delta v = \dot{\omega}t^2/2$.

---

[3] Compare with $R = \exp(\tilde{r}) = (I + \tilde{r}) \cdot (I - \tilde{r})^{-1}$ (Cayley formula).

In the case of a combination of motion one must express these quantities in a fixed frame of reference and then perform the right frame transform.

Reciprocally, the angular velocity $\boldsymbol{\omega}$ can be related to the infinitesimal instantaneous rotation by (details are in [4.2], pp. 215–217):

$$\frac{\mathrm{d}R}{\mathrm{d}t}R^{\mathrm{T}} \cdot \boldsymbol{x} = \tilde{\omega} \cdot \boldsymbol{x} = \boldsymbol{\omega} \wedge \boldsymbol{x} \ .$$

This defines a linear homogeneous differential equation with non-constant coefficients for $R(t)$. It is well known [4.11] that *such an equation has a unique solution, function of the initial value $R(0)$*. This result is to be interpreted as follow: "One cannot define absolute angular orientation, but only relative angular orientation with respect to a reference". Then, an initial orientation being given, the angular orientation can be expressed as *the rotation to be performed to get from the initial orientation to the actual orientation*. There is no explicit solution for such an equation in the general case.[4]

This numerical integration is, however, particular. Let us note for example

$$\Delta R(t) = 1 + \tilde{\omega}(t)$$

which represent an infinitesimal rotation, we must, in the case of a discrete formulation, combine these rotations. We have:

$$R((n + 1) \cdot \Delta t) = \Delta R(n \cdot \Delta t) \cdot R(n \cdot \Delta t) = \prod_{p=0}^{n} [1 + \tilde{\omega}(p \cdot \Delta t)] \ .$$

This combination is *not additive* but multiplicative, and the product is *not commutative*.[5] In addition, because of numerical errors, this product of orthogonal matrixes generates a non-orthogonal matrix, and this computation *is not stable*.

In fact, combining two general rotations is then a complicated problem solved by the *Campbell and Hausdorff* formula [4.12]. Using previous notations, a rotation can be expressed as

$$R = \mathrm{e}^{\tilde{r}}$$

where the exponential of matrix is used here. The direction of the rotation is given by the direction of the vector $\boldsymbol{r}$, while the angle of the rotation, taken as positive, is the norm of the vector $\boldsymbol{r}$. In our case we have

---

[4] There is an explicit solution in the case where
$$\forall t_1, t_2 \ , \ \tilde{\omega}(t_1) \text{ commutes with } \tilde{\omega}(t_2) \ .$$
This is true if and only if the rotation is performed around a fixed axis, that is if $\boldsymbol{\omega}(t) = \theta(t) \cdot \boldsymbol{u}$, with $\boldsymbol{u}$ fixed. In this case we get
$$R(t) = \exp\left\{ \left[\int_0^t \theta(t) \, \mathrm{d}t \right] \cdot \tilde{u} \right\} \cdot R(0) \ .$$

[5] One might wish to take only the first terms of this last product, and use
$$R((n + 1) \cdot \Delta t) \simeq 1 + \sum_{p=0}^{n} \tilde{\omega}(p \cdot \Delta t) + o(\| \tilde{\omega}(t)\|^2)$$
which is additive and commutative in terms of $\tilde{\omega}(t)$. Such formula give bad results, since the matrix $R(t)$ computed with such a formula is not orthogonal because (among other things) its diagonal coefficients are equal to one. An improved method has therefore to be used.

$$R(t + \Delta t) \quad = \quad e^{\tilde{r}(t+\Delta t)} = e^{\tilde{\omega}(t)} \cdot e^{\tilde{r}(t)} \;,$$
$$1 + \tilde{\omega}(t) \quad \simeq \quad e^{\tilde{\omega}(t)} \;.$$

The *Campbell and Hausdorff* formula gives an explicit expression of the exponential of the product of two exponentials in a Lie-group, in term of a series. In our case, using the third order of the development of this series of matrix, we have

$$\exp\left(\tilde{r}(t + \Delta t)\right) \quad \simeq \quad \exp\left(\tilde{\omega}(t) + \tilde{r}(t)\right) + \frac{1}{2} \cdot [\tilde{\omega}(t), \tilde{r}(t)]$$
$$+ \frac{1}{12} \cdot \left([\tilde{\omega}(t), [\tilde{\omega}(t), \tilde{r}(t)]] + [\tilde{r}(t), [\tilde{r}(t), \tilde{\omega}(t)]]\right)$$

which is valid since the operation:

$$[\tilde{x}, \tilde{y}] = \tilde{x} \cdot \tilde{y} - \tilde{y} \cdot \tilde{x}$$

transforms two 3D antisymmetric matrices into one 3D antisymmetric matrix, which can be associated to the skew of a 3D vector.

This method is the only one method which allows a direct computation of the combination of two rotations. The representation of the rotation is minimal (3 parameters), and the operation directly maps two rotations onto a rotation, without the necessity of fining the closest unitary element (as it is the case for matrix and quaternions). This operation is a polynomial function of the components of the rotation vectors.

However this method has two drawbacks. First, the number of operations is very high. For example, for the 3rd-order expansion, we have

$$\left|\begin{array}{l} \omega_x + r_x + \dfrac{\omega_y r_z}{2} - \dfrac{\omega_z r_y}{2} - \dfrac{\omega_x r_y^2}{12} + \dfrac{\omega_y r_x r_y}{12} + \dfrac{r_z \omega_z r_x}{12} - \dfrac{\omega_x r_z^2}{12} \\ \omega_y + r_y + \dfrac{\omega_z r_x}{2} - \dfrac{\omega_x r_z}{2} - \dfrac{\omega_y r_z^2}{12} + \dfrac{r_z \omega_z r_y}{12} + \dfrac{\omega_x r_y r_x}{12} - \dfrac{\omega_y r_x^2}{12} \\ \omega_z + r_z + \dfrac{\omega_x r_y}{2} - \dfrac{\omega_y r_x}{2} - \dfrac{\omega_z r_x^2}{12} + \dfrac{\omega_x r_z r_x}{12} + \dfrac{r_y \omega_y r_z}{12} - \dfrac{\omega_z r_y^2}{12} \end{array}\right.$$

which is computed using 21 additions and 45 multiplications. Second, this is only the 3rd-order expansion, and the calculation has to be carried on, since r corresponds to the angular position, and is *a priori* small.

4.5.3 Cooperation of the Inertial and Visual Systems

Let us now introduce the use of inertial information in a visual system. Inertial information provides a partial estimation of the **robot self-motion** and the **robot absolute orientation**, precisely **angular velocity** and **specific forces**.

Although only the gravity, the rigid body acceleration, and the angular instantaneous velocity can be measured, it is possible to derive from these quantities the following parameters: **Instantaneous angular velocity**, which is directly given by the gyrometers output. **Linear acceleration**, which is not

directly output by the accelerometers but added to the gravity. There is
no possibility, *a priori* to separate linear acceleration and gravity, while additional internal hypotheses will be used. **Vertical orientation**, which is
directly given by the direction of the gravity vector, and which is available as
soon as linear acceleration and gravity are differentiated. **Angular position**,
which is computed by the integration of gyrometers output, or measured using gyroscopes. Angular position evaluation is subject to a drift, linear with
respect the time, and has to be periodically re-calibrated. **Linear translation**, which is computed by the integration of linear acceleration, and is also
subject to a drift.

Internal representations of instantaneous self-motion and angular orientation can be based on these parameters. We are going to analyse how such
quantities can be computed from sensors output, and study the use of these
quantities in cooperation with odometric and visual information.

4.5.4 Application of Inertial Information in a Visual System

Visual information also provides information about motion and orientation.
In order to cooperate with inertial information, either a differential approach
(like "optical flow") of visual information, cr a "token-tracker" based approach analysis is used.

Cooperation between visual and inertial information can be:

Estimation of instantaneous motion. In a rigid, almost stationary background the linear velocity (up to a scale factor) and the angular velocity can
be estimated, from the image irradiance variations, on pixels with high contrast. Combined with the inertial estimations of angular velocity and linear
velocity time derivative, this estimation can be improved and sped-up.

Estimation of angular orientation. Orientation with respect to the vertical
is provided by inertial cues while the variation of orientation in the horizontal
plane can be integrated using visual cues, providing a robust estimation of
absolute angular orientation.

3D Vision. When motion is known, the structure of the environment can
be inferred from visual information. Using inertial cues in order to estimate
motion, it is possible to build an instantaneous depth map of the environment.
Using several frames the precision of this estimation is improved.

Intrinsic calibration of the visual sensor. In the case of rotations around
the optical center of the visual sensor, the variations of irradiance are no
longer related to the environment structure, but only depend on the sensor
geometry. Using inertial cues, it is possible to generate such rotations and to
perform the visual sensor calibration in any, structured, visual environment.

Segmentation between a moving object and the background. The background being stationary and its relative movement with respect to the visual
sensor being known through visual cues, it is possible to determine which
part of the image belongs to a moving object, the segmentation being based
on motion disparity.

All of these applications require an estimation of the angular orientation and of the linear velocity time derivative, computed from sensors output.

4.5.5 Computation of the Scale Factor

Moreover we can solve the well known "scale factor problem" using inertial cues. Using monocular sequences of images, and because of the nature of the visual projection, it is not possible to recover the entire 3D structure. On the contrary, every thing is known only up to a scale factor. Roughly speaking, a small object moving slowly and close to the retina produces the same visual perception as a bigger object moving faster, but further away from the retina.

This means that monocular visual cues do not recover the linear velocity v, but v up to a scale factor (focus of expansion), say $f = v/||v||$, whereas inertial cues do not recover v, but its derivative $\gamma = \dot{v}$.

Now, let us write $v = vf$ with $||f|| = 1$, and $v = ||v||$.

We have $\gamma = \dot{v}f + v\dot{f}$ and since $f \perp \dot{f}$ because f is unitary:

$$||\dot{f}|| \neq 0 \quad \Rightarrow \quad v = \frac{\gamma^{\mathrm{T}} \cdot \dot{f}}{||\dot{f}||^2} .$$

We can thus compute the norm of the linear velocity, and obtain a complete estimate of the linear velocity by a cooperation between visual and inertial cues. Thus we can use a monocular sequence of images and avoid stereo.

4.6 Computing Self-Motion with a Vertical Estimate

Let us define a representation of self-motion in which the following "natural" notions are explicit: 3D vertical, height and rotation around a fixed axis.

Any robotic system, on earth, is embedded in the earth's gravity field. As a result, the measurement of the absolute vertical is possible using low-cost inertial sensors [4.6]. The key point is that the gravity acceleration is a constant, homogeneous vector. It's orientation is a very precise indicator of absolute vertical axis. The vertical thus constitutes a basic cue for spatial orientation. Using this, we can relate our visual information to an absolute frame of reference, with the vertical as a fixed axis.

Moreover, it is possible to compute the vertical direction in a natural visual scene, since there often are vertical lines that intersect at infinity are define a vanishing point, the projection of which can be detected in the picture [4.13]. In addition, on a mobile robot or for an industrial mount, the approximate orientation of the camera is known. This allows us to compute an approximate location of the vertical direction.

Knowing the vertical, we will show in this book that it is possible to reproject the visual tokens in such a way that the vertical in the image is aligned with the true 3D vertical. In such a case, the rotation of the visual token reduces to a 2D rotation in the horizontal plane, much simpler to compute.

Finally, it is possible to sort the line segments, depending whether they are vertical or not. In an indoor or outdoor scene such tokens are common (trees, buildings, indoor walls, posts, doors or windows edges, etc.). For vertical line segments, 3D orientation is indeed known.

An estimate of the vertical can be obtained in three situations: One has a coarse estimate of the vertical, and there exist vertical lines in the image; one has an accurate estimate of the vertical, from odometric or inertial cues; the visual system moves in a plane parallel to the floor (mobile robot), or is fixed (robotic arm) with respect to the environment, and thus the vertical is known from calibration.

One of the above conditions is almost always true, and considering artificial visual systems mounted on robots, this approach is quite general.

Let us say, from now, that g *is aligned with the 3D vertical*, when considering an absolute 3D frame of reference, while "vertical in the image" is aligned with y.

4.6.1 A Representation of Self-Motion Using Vertical Cues

In an environment related to a vertical, motion must not be analysed in a homogeneous way. Such "non-homogeneous" approachs have been proposed long ago [4.14]. In our case, we have this notion of "orientation with respect to the vertical". The rotation is to be decoupled into two components: a vertical rotation around a fixed vertical axis and *gyro-rotation*, which corresponds to the rotation that brings the system into alignment with the vertical. The translation is also not analysed in an homogeneous way. On the contrary, we would like to introduce the notion of "motion in height", and horizontal motion. This is a very natural distinction in the case of a mobile robot, or for a robotic system aboard any vehicle. In particular, assuming the system is moving on a horizontal plane, motion in height is expected to be null. In any case, this distinction is to be made if we want to decompose the motion in two components: one in the vertical direction and one in the horizontal plane.

When using inertial cues, the quality of the signal is dependent upon the direction of motion, since the system is working with a important offset in the vertical direction. Specifically, during a motion in the vertical direction, the related acceleration is small with respect to gravity, and thus vertical accelerations are very small with respect to the signal average and are almost taken as noise. In the horizontal directions, the same acceleration is measured without any offset, resulting in better measurements.

A step further, translational motion in height is a pure translation, while translational motion in the horizontal plane might be interpreted in two ways: either as a true translation, or in combination with rotation in the horizontal plane as a rotation around an axis off-centered with respect to the origin.[6] In order to take into account the fact that rotation and translation performed in the horizontal plane are not independent we are going to consider a pure off-center rotation, around a vertical axis.

Let us now formalize the previous discussion. If the camera is calibrated, then the geometric quality of actual CCD sensors legitimates the use of the pinhole model for the camera, with a unit focal distance. Every quantity is referred to in the camera intrinsic coordinates. The origin of this frame of reference is the optical center of the camera (its image nodal point, in fact) and the Z axis corresponds to the optical axis of the camera.

All coordinates are related to a frame of reference attached to the retina (x, y, z), z being aligned with the optical axis, y being aligned with the vertical in the image.

We consider a 3D point $M = (X, Y, Z, 1)^T$ in homogeneous coordinates and its projection $m = (x, y, 1)^T$ in the image plane. For a rigid motion we have $M' = R(u, \theta) \cdot M + t$ where $R(u, \theta)$ is the rotation matrix and t the translation vector.[7] More precisely we can write

$$m = \begin{pmatrix} 1 & 0 & 0 & 0 \\ 0 & 1 & 0 & 0 \\ 0 & 0 & 1 & 0 \end{pmatrix} \cdot M' = P \cdot M',$$

$$m \equiv P \cdot [R(u, \theta) \cdot M + t] . \tag{4.13}$$

Similarly, a line direction (vanishing point) $L = (X, Y, Z, 0)^T$ and its projection $l = (x, y, l)^T$ are related by

$$l = P \cdot L',$$
$$l = P \cdot [R(u, \theta) \cdot L] . \tag{4.14}$$

We consider that the rigid motion is decomposed into three components:

A rotation around a 3D vertical axis aligned with g. This rotation is not necessarily performed around the origin where the axis is intersecting the horizontal plane at the point $C = (C_x, 0, C_z)^T$. Its angle is noted $\Theta \in]-\pi, \pi[$. Note that we exclude the symmetries in the horizontal plane, that is rotations around a 3D vertical axis aligned with g with angle π.

A translation $T = (0, h, 0)^T$ in height, in the direction of the 3D vertical.

A "gyro-rotation": a rotation around an axis in the horizontal plane which allows the 3D vertical to be aligned with the vertical of the image. The rotation axis is parameterized by $\Omega = (\cos(\rho), 0, \sin(\rho))^T$ with $\rho \in]-\pi, \pi]$, and the rotation angle by $\mu \in [0, \pi[$.

Note that we exclude again the symmetries (rotations of angle π).

---

[6] One should remember that an angular rotation ω around a point C induces a linear translation at a point M equal to $\omega \wedge CM$.

[7] In this book, scalars are noted using normal font, vectors using bold fonts, and matrix using capital letters. The sign $\wedge$ is used for cross-product and the sign $\cdot$ for the matrix product. As a consequence the dot-product of two vectors is denoted $\langle u, v \rangle = u^T \cdot v$.

Using our representation we have:

$$
M' = \begin{pmatrix} & & & 0 \\ & R(\boldsymbol{\Omega}, \mu) & & 0 \\ & & & 0 \\ 0 & 0 & 0 & 1 \end{pmatrix} \cdot \begin{pmatrix} & & & 0 \\ & Id_{3\times3} & & h \\ & & & 0 \\ 0 & 0 & 0 & 1 \end{pmatrix}
$$

$$
\cdot \begin{pmatrix} & & & 0 \\ & R(\boldsymbol{y}, \Theta) & & 0 \\ & & & 0 \\ 0 & 0 & 0 & 1 \end{pmatrix} \cdot \left(M - \begin{vmatrix} C_x \\ 0 \\ C_z \end{vmatrix} \right) \tag{4.15}
$$

for a point, and a similar relation for a line direction.

The notation $R(\boldsymbol{u}, \theta)$ corresponds to a compact representation of the rotations discussed previously.

Proposition 4.2. The six parameters $(\rho, \mu, h, \Theta, C_x, C_z)$ form a minimal, one to one, local representation of the rigid motion $M' = R(\boldsymbol{u}, \theta) \cdot M + \boldsymbol{t}$.

The representation is non-singular except if the rotation angle is zero or if one of the rotation is of angle π. These singularities are simple.

Explicit formulas exist between this representation and usual representations of rigid motion.

Proof. The representation is minimal since rigid motion is a six dimensional manifold.

Expanding (4.15) we obtain

$$
R(\boldsymbol{u}, \theta) = R(\boldsymbol{\Omega}, \mu) \cdot R(\boldsymbol{y}, \Theta) ,
$$

$$
\boldsymbol{t} = R(\boldsymbol{\Omega}, \mu) \cdot \left(\begin{vmatrix} 0 \\ h \\ 0 \end{vmatrix} - R(\boldsymbol{y}, \Theta) \cdot \begin{vmatrix} C_x \\ 0 \\ C_z \end{vmatrix} \right) ;
$$

this yields:

$$
\sin(\theta)\boldsymbol{u} = r(\rho, \mu, \Theta) = \frac{1}{2} \begin{vmatrix} \sin(\mu) \left[\cos(\rho) + \cos(\rho + \Theta) \right] \\ \sin(\Theta) \left[1 + \cos(\mu) \right] \\ \sin(\mu) \left[\sin(\rho) + \sin(\rho + \Theta) \right] \end{vmatrix} .
$$

Now recalling that the antisymmetric part of the rotation is simply

$$
\frac{R(\boldsymbol{u}, \theta) - R(\boldsymbol{u}, \theta)^{\mathrm{T}}}{2} = \sin(\theta)\tilde{\boldsymbol{u}}
$$

we have

$$
\|r(\rho, \mu, \Theta)\|^2 = \frac{1}{4} \left[1 + \cos(\mu) \right] \left[1 + \cos(\Theta) \right]
$$

$$
\cdot \left[3 - \cos(\Theta) \cos(\mu) - \cos(\mu) - \cos(\Theta) \right] .
$$

Then considering $\Theta \in] - \pi, \pi[$, $\rho \in] - \pi, \pi]$ and $\mu \in [0, \pi[$ as in the definition, we have a mapping from our representation to usual representation of 3D rotations, since we can compute $\boldsymbol{u} = r(\rho, \mu, \Theta)/\|r(\rho, \mu, \Theta)\|$ and $\theta = \arcsin(\|r(\rho, \mu, \Theta)\|)$. This mapping becomes singular only if $\theta = 0$.

In the reverse direction, we can explicitly state the relations, using the rotation matrix:

$$\rho = \arctan\left[-\frac{R(u,\theta)_{xy}}{R(u,\theta)_{zy}}\right] ,$$

$$\mu = \arccos\left[R(u,\theta)_{yy}\right] ,$$

$$\Theta = -\rho + \arctan\left[\frac{R(u,\theta)_{yx}}{R(u,\theta)_{yz}}\right] .$$

This provides a mapping from the usual representation to ours.

These inverse trigonometric relations hold for the intervals the angles are defined in, and are defined for all rotations except these for which $R(u,\theta)_{xy} = R(u,\theta)_{zy} = 0$ (pure vertical rotations) or $R(u,\theta)_{yx} = R(u,\theta)_{yz} = 0$ (pure horizontal rotations). In the first case ρ is indeed undefined since it is related to the orientation of a zero angle rotation, whereas in the other case we simply have $\Theta = 0$, this last case being thus not a new singularity.

On the other hand

$$t = R(u,\theta) \cdot \begin{vmatrix} -C_x \\ h \\ -C_z \end{vmatrix}$$

since $\begin{vmatrix} 0 \\ h \\ 0 \end{vmatrix}$ is invariant through $R(y,\Theta)$. The parameters of the translation are thus related by a simple non-singular linear relation.

We thus have a complete one to one mapping between the two representations.   □

4.6.2 The "Gyro-Rotation" and the Vertical Rectification

After a rigid motion, neither the translation nor the vertical rotation, but only the gyro-rotation affects the orientation of the vertical. More precisely, if g is a vector aligned with the vertical we have after a rigid motion as defined in (4.15):

$$\frac{g}{\|g\|} = \begin{vmatrix} -\sin(\mu)\sin(\rho) \\ \cos(\mu) \\ \sin(\mu)\cos(\rho) \end{vmatrix} . \tag{4.16}$$

It is thus straightforward to compute the characteristics of the "gyro-rotation" from the vertical orientation:

$$\rho = -\arctan(g_x/g_z) , \quad \mu = \arctan\left[\sqrt{(g_x^2 + g_z^2)}/g_y\right] . \tag{4.17}$$

Now these two sets of equations provide a one-to-one mapping between the direction of a any vertical vector g and the gyro-rotation parameters

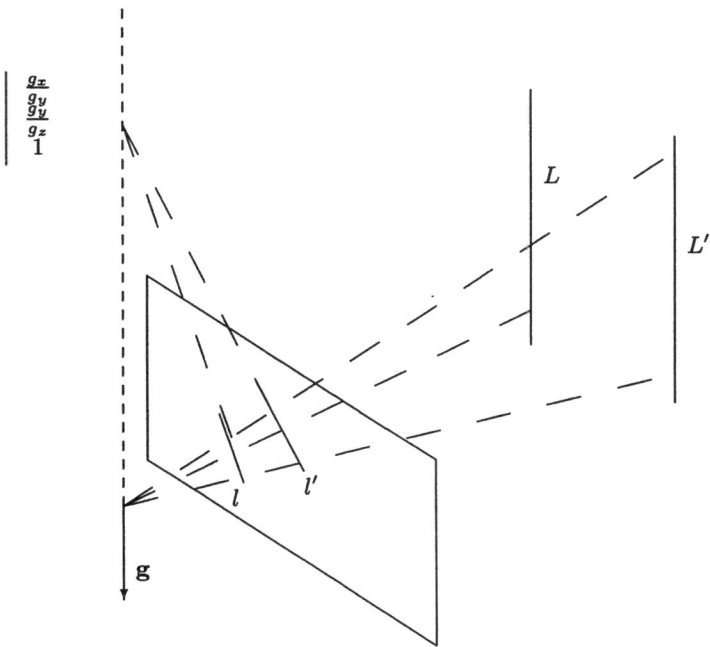

Fig. 4.4. Vanishing point corresponding to vertical lines

(ρ, μ). This mapping is singular only when the vertical vector is null or if the gyro-rotation is null.

Moreover, let us consider the projection defined in (4.13). Vertical lines in space will project onto lines in the image plane and intersect in a "vanishing point" corresponding to the projection of the vertical point at infinity in space, as illustrated in Fig. 4.4.[8]

Now, we also have

$$
P \cdot \begin{pmatrix} & & & 0 \\ & R(\Omega, \mu) & & 0 \\ & & & 0 \\ 0 & 0 & 0 & 1 \end{pmatrix} = R(\Omega, \mu) \cdot P \, ,
$$

---

[8] To demonstrate this, we give a simple example: let $(X, Y, Z)^{\mathrm{T}} + \lambda g$ be a general point that lies on a vertical line through any point $(X, Y, Z)^{\mathrm{T}}$. The related projection is a point having $\alpha(X, Y, Z)^{\mathrm{T}} + \beta g$ as homogeneous coordinates, thus defining a projected line. There is a common point for this pencil of lines, with homogeneous coordinates $(g_x, g_y, g_z)^{\mathrm{T}}$ (take $\alpha = 0$ and $\beta = 1$ for all lines). But now, computing the transformation of g through a gyro-rotation $R(\Omega, \mu)$ as defined in (4.17), yields: $R(\Omega, \mu) \cdot g = (0, 0, 1)^{\mathrm{T}}$. This last vector corresponds to the vertical direction in the image plane, thus establishing the relation between g and the vanishing point.

the rotation in the right hand size being a homogeneous transformation of the image points. *It is possible to reproject the image in such a way that there is no more gyro-rotation, while the vertical columns of the image are aligned with the absolute vertical.*

All this can be summarized in the following proposition:

Proposition 4.3. Given an estimate of the 3D vertical it is possible to compute the unique gyro-rotation corresponding to a relative motion for which the vertical axis of the system will be aligned with the rows of the image frame.

The reprojection matrix is precisely $R(\boldsymbol{\Omega}, \mu)$.

Moreover, among all rotations which aligns the vertical in space with the vertical in the image $R(\boldsymbol{\Omega}, \mu)$ is the one with the smallest angle.

Proof. Consider a rotation matrix $R(\boldsymbol{u}, \Theta)$ with $||\boldsymbol{u}||$. This aligns the vertical in space with the vertical in the image if $R(\boldsymbol{u}, \Theta) \cdot \boldsymbol{y} = \boldsymbol{g}$.

Compute $\boldsymbol{y}^{\mathrm{T}} \cdot \boldsymbol{g} = \boldsymbol{y}^{\mathrm{T}} \cdot R(\boldsymbol{u}, \Theta) \cdot \boldsymbol{y}$ and simplify. We obtain

$$\boldsymbol{y}^{\mathrm{T}} \cdot \boldsymbol{g} = [1 - \cos(\Theta)] \, \boldsymbol{u}^{\mathrm{T}} \cdot \boldsymbol{y} + \cos(\Theta)$$

yielding

$$\cos(\Theta) = \frac{\boldsymbol{y}^{\mathrm{T}} \cdot \boldsymbol{g} - \boldsymbol{u}^{\mathrm{T}} \cdot \boldsymbol{y}}{1 - \boldsymbol{u}^{\mathrm{T}} \cdot \boldsymbol{y}} .$$

It is clear from this equation that Θ is minimum if and only if $\boldsymbol{u}^{\mathrm{T}} \cdot \boldsymbol{y} = 0$, that is, if the axis of rotation is orthogonal to $\boldsymbol{y}$ and thus in the horizontal plane, as claimed. □

Applying this gyro-rotation matrix to the image will be called a *vertical rectification* since it is similar to a stereo rectification transforming epipolar lines to retinal horizontal lines. In fact, for a stereo algorithm, the stereo geometry is simplified by a vertical rectification since both image planes are vertical.[9]

This vertical rectification can be applied on all features which are projectively invariant: points, lines or ellipses. Only the first two primitives are taken into account here, but generalization to the third one is straightforward [4.15].

In an image sequence, **tracking** is to be performed after image rectification. This intermediate process minimizes the token relative displacements between two frames and simplifies the tracking phase, because it reduces the disparity between two related tokens and thus simplifies the matching to be performed by the token-tracker [4.16].

The estimate of the vertical from inertial cues has been already discussed, and so we now discuss how to compute the vertical from visual cues.

---

[9] In such a system, stereo has to be performed after all the preprocessing described in this section and the next one. Each token will have a 3D representation, thus simplifying the correspondence problem, and the stereo problem will be then a "sensor fusion" problem.

Correcting the Vertical Estimate from Visual Cues. The computation of vanishing points is very common in vision, and is in general an ill-conditioned problem. However, in most studies [4.13], the fact that an *a-priori* estimate of the vertical might be available is not taken into account.

This is, however, the case in practical context and we will use this fact now. More precisely, given a vertical estimate g, any vertical line should pass through the 2D point $\mathcal{G} = (g_x/g_z, g_y/g_z, 1)$ using homogeneous coordinates. When this is not the case, a natural idea is to consider that *the "conditional vertical estimate" given by a projected vertical line and a known coarse estimate of the vertical, is the projection of the vertical vanishing point onto the projected vertical line.*

Using this simple property (proof left to the reader) we have:

Proposition 4.4. *The projection of a point* $g = (g_x/g_z, g_y/g_z, 1)$ *onto a line* D *with equation* $\cos(\theta)x + \sin(\theta)y - c = 0$ *is in homogeneous coordinates:*

$$g_D = \begin{vmatrix} x \\ y \\ u \end{vmatrix} = \left(\begin{array}{ccc} \sin(\theta)^2 & -\cos(\theta)\sin(\theta) & c\cos(\theta) \\ -\cos(\theta)\sin(\theta) & \cos(\theta)^2 & c\sin(\theta) \\ 0 & 0 & 1 \end{array} \right) \cdot \begin{vmatrix} g_x \\ g_y \\ g_z \end{vmatrix}.$$

Given a vertical direction g, we can thus estimate for each projected line its conditional vertical estimate g_D. The vector from the estimate onto the line to the coarse estimate will be considered as an estimate of the error for this line. The covariance of this estimate is easy to obtain from first order approximation, a function of the covariance on g and the variance of θ, both quantities being taken as independent.

Given a set of lines almost vertical and a coarse estimate of the vertical, it is thus possible to have a conditional estimate and an evaluation of the bias for each measurements [4.17].

The filter assigns to each segment a weight which is proportional to the distance between the vanishing point and its line support. The threshold on this value allows the algorithm to sort vertical and oblique segments with respect to the accuracy.

This algorithm will not only estimate the vertical but also sort the lines in term of vertical or non-vertical lines.

4.6.3 Stabilization Using 2D Translation and Rotation

Rectification of a 2D Stabilized Picture. The gyro-stabilization developed in the last section provides pictures whose verticals are aligned with the 3D vertical.

We now want to complete the stabilization of the images sequence, determining the 2D linear translation and polar translation[10] which minimize the disparity between two successive views.

---

[10] By "polar translation" we mean a collineation of the plane that translates the orientation of some lines. This polarity might be a simple rotation around, say, the projection of the optical center. But not always. If it modifies the points at infinity, as used here, it is not a Euclidian rotation but a true collineation of the image plane.

This is related to the three parameters: a horizontal translation du, a vertical translation dv, and a polar translation around the optical center with angle $d\theta$.

As in the previous section, for a given set of parameters $(du, dv, d\theta)$, we would like to compute a collineation of the plane, corresponding to a 3D rotation which cancels the horizontal, vertical and polar disparities. Such a rotation R_S must verify the following two constraints:

$$\begin{pmatrix} 0 \\ 0 \\ 1 \end{pmatrix} = R_S \cdot \begin{pmatrix} sdu \\ sdv \\ s \end{pmatrix} \qquad \begin{pmatrix} 0 \\ 1 \\ 0 \end{pmatrix} = R_S \cdot \begin{pmatrix} s\sin(\theta) \\ s\cos(\theta) \\ s\mu \end{pmatrix}. \qquad (4.18)$$

The first constraint corresponds to the cancellation of the 2D translation $(du, dv)^\mathrm{T}$; with this equation, the displaced origin is put at $(0, 0)$.

The second constraint corresponds to the polar stabilization. With this equation, any point on a line with vertical orientation θ is put onto a true vertical line of the image. Please note that this does not corresponds to a true 2D rotation, but a polar collineation of the plane, since we allow a point at finite distance to be mapped onto the vanishing point.

Let us explain why. Considering a true rotation corresponds to having $\mu = 0$ in the previous equation, since the vertical vanishing point $(0, 1, 0)$ must be mapped onto another vanishing point. In such a case, we have

$$\begin{pmatrix} s'\sin(\theta) \\ s'\cos(\theta) \\ 0 \end{pmatrix}^\mathrm{T} \cdot \begin{pmatrix} sdu \\ sdv \\ s \end{pmatrix} = ss' \left[du\sin(\theta) + dv\cos(\theta) \right] = 0$$

since these two vectors transform to orthogonal vectors by the rotation. This is not always the case, whereas there is a constraint between the 2D linear translation and the rotation angle. In other words, in order to be capable of defining an orthogonal homogeneous matrix which translates and "rotates" the plane to cancel the 2D disparity, we must "reproject" the picture.

The reader can easily verify that there is only one orthogonal homogeneous matrix which verifies (4.18). Its explicit form is easy to obtain.

This is coherent with the fact a rotation matrix is defined by three parameters, while (4.18) correspond to three constraints.

Computing the Translational and Polar Disparity. Let us make the hypothesis of relatively small displacements between two views. This is legitimated by the high frequency used during the acquisition of the images sequence

One idea to compute the horizontal and vertical disparity between two frames is to observe the distribution of the abscissa of and ordinates for the point belonging to the line segments detected on the image. We can easily compute the histograms corresponding to these distributions.

By making the histograms correspond to the horizontal distribution in two consecutive frames, we can shift them and take the value which maximizes the correlation between them. This value of disparity is a possible estimate of the horizontal disparity du.

A similar method has to be applied on the ordinates to compute dv, and to the line segment orientations to obtain an estimate of $d\theta$. However, since we already stabilized the picture through a gyro-rotation, we did not have to search for a polar disparity, expected to be zero, and simply fixed $\theta = 0$.

We thus have to calculate two histograms for each image which correspond to a projection of the line segments onto the axes. We must avoid the border of the image where some segments may appear or disappear. After several experimentations, we found a reasonable criterion to fill the histograms: Each slot of the histogram is incremented by a value which is given by the following formula:

$$\text{Histo}\,[k] + = \left(\frac{2 \, \text{length} \, \pi}{x_2 - x_1} \right) \cos \left[\frac{\pi \, (2k - x_1 - x_2)}{x_2 - x_1} \right] .$$

In the previous formula x_2 and x_1 are the abscissa or ordinates of the 2D segment extremities. The contribution to the histogram is null at the extremities of the segment, and maximal at the middle of the segment. We multiply the cosine function by a normalization coefficient in order to have its integral value equal to segment length.

Having these histograms for two successive images, we search to maximize their correlation. The maximum of the correlation gives us the best 2D translation as expected. According to the first hypothesis, we have limited the shifting between the histograms when calculating the correlation to ± 30 pixels. We have chosen this method, which gave us good results as shown by our experimental results.

4.6.4 Implications on the Structure from Motion Paradigm

Let us consider the rotation matrix $R^* = R_S \cdot R(\mathbf{\Omega}, \mu)$. This matrix, computed in two steps, corresponds to a transformation of the plane that reduces the disparity between line segments. If the true 3D rigid motion is a rotation, or if the translation is small, this should provide (1) a fairly good estimate of the rotational self-motion and (2) cancel the token disparities and simplify the correspondence problem.

4.6.5 Estimation of the 3D Rotation

Let us assume our estimate of the 3D rotation is R^*. Here, we obtain explicit values for the 2D motion components, instead of implicit nonlinear equations. We thus avoid the use of nonlinear estimation techniques. In other words, we prefer to control the approximations to be made, rather than writing true but "unsolvable" equations that are processed blindly by nonlinear minimization routines.

In the case of lines, the general method requires matchings between tokens in three frames [4.18], thus yielding heavy computations and tracking problems. Our computations are done without correspondences.

However, the estimated rotation R^* might differ from the true rotation R for two reasons.

First, the estimation of translational and polar disparities is done on the image plane and this value does not correspond in a given location of the image to the disparity obtained by a 3D rotation. Due to the definition of the stabilization, it does so only at the origin. However, if the disparity is zero, the solution is $R = R^* = 0$. Then, the smaller the disparity, the smaller the approximation. It is then possible to overcome this bias by an iterative process that computes an estimate of the disparity, reprojects, recomputes an estimate of the reduced disparity, reprojects again and so on. Such a process converges if and only if the disparity is reduced at each step.

The second reason why R^* might differ from R is that, the motion of visual tokens is not only function of the rotational motion but also of the 3D translation. Exception are points or lines at the horizon, that is, points or lines with a depth much higher than the translation. More precisely, we have $Z'm' = ZR \cdot m + t \simeq ZR \cdot m$, if Z is high. The projection of such a point is only function of the rotational self-motion. Then if R^* is close to the true rotation, points or lines which 2D motion is almost cancelled by R^* will thus correspond to points or lines at horizon (i.e. with an important depth), whereas points or lines which still have a disparity after a reprojection using R^* will correspond to tokens with an *observable depth*, or with another rigid motion.

It is therefore possible to detect points at the horizon by the observation of this disparity.

4.6.6 Implication on Token Matching

Since the average disparity between two line segments is expected to be small, the correspondence problem might be simply solved, avoiding the use of a complex token tracker. Consider a set of n_0 line segments in one frame and n_1 segments in the previous frame, and compute the $n_0 n_1$ *Mahalanobis distance* between these pairs of segments. Such a distance is easily computed, considering, for instance, the middle points $m_{\bullet 0}$ and $m_{\bullet 1}$ and the orientations θ_0 and θ_1 of each segment [4.16, 19]. In that case, the distance has the explicit form

$$\delta_{01} = (m_{\bullet 0} - m_{\bullet 1})^{\mathrm{T}} \cdot (\Lambda_{m\bullet 0} + \Lambda_{m\bullet 1})^{-1} \cdot (m_{\bullet 0} - m_{\bullet 1}) + \frac{(\theta_0 - \theta_1)^2}{V_{\theta_0} + V_{\theta_1}}$$

where $\Lambda_{m\bullet 0}$ is the covariance on the middle point and V_{θ_0} the variance on the line segment orientation.

After computing these $n_0 n_1$ values, we can perform the matching by first choosing the pair of segment which have the smallest distance, and then removing these segments from the segment lists. This operation is repeated until the distances between two segments are higher than a given threshold, corresponding to segments which are not likely to match.

4.6.7 Experimental Results

In order to show the usefulness of our approach, we have applied this method on various scenes, including scenes in which line segments are not so easy to detect as shown in Fig. 4.5. We thus are detected "microscopic" line segments on each point of the edges. Very simply, each line-segment were defined as tangent to the edge using the gradient orientation.

Fig. 4.5. Using a scene with an object in motion (*the white box*)

The mechanism of stabilization has been checked when a moving object was present in the scene, as shown in Fig. 4.6 and Fig. 4.7. It is clear from this result that the self-motion has been canceled whereas the object motion is much more visible. Detected line-segments were very small here, as visible on the display.

Quantitatively, the vertical lines orientation is better than 0.2 deg (standard deviation).

The mechanism of stabilization has been checked also on outdoors scene as shown in Fig. 4.8 and Fig. 4.9. It is clear from this result that vertical stabilization is still working on this kind of scene. Moreover, it is visible that

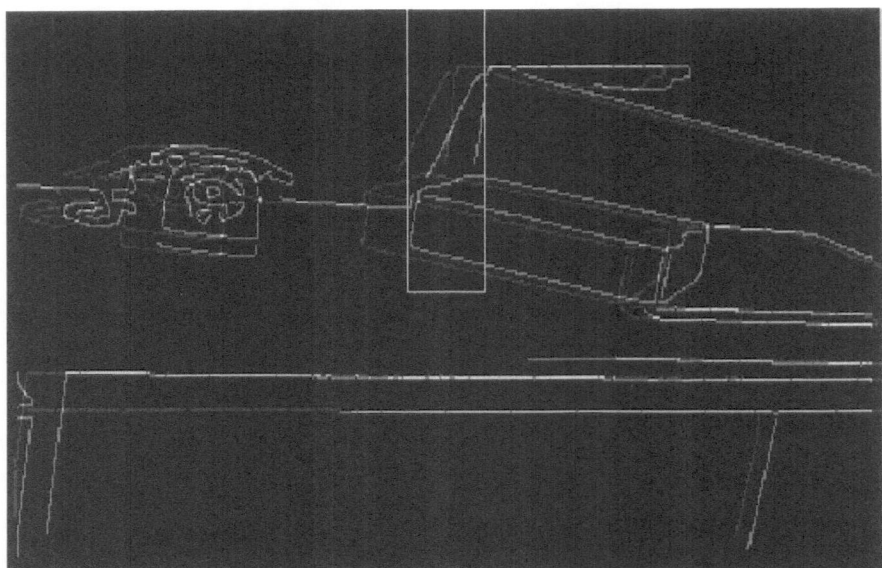

Fig. 4.6. Two consecutive views with a moving object displayed before stabilization

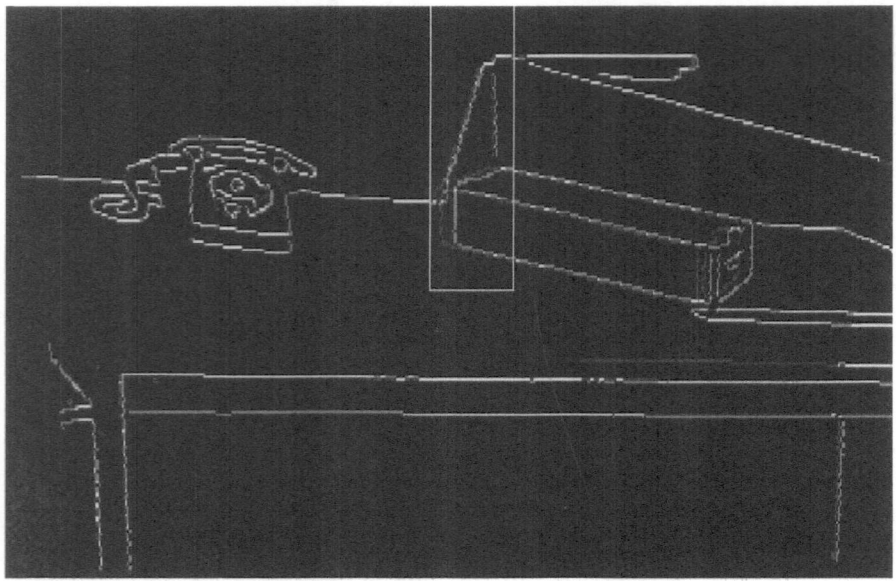

Fig. 4.7. Two consecutive views with a moving object displayed after stabilization

remote objects are better stabilized than closer objects, as expected. The corresponding histogram of is shown in Fig. 4.10.

Fig. 4.8. Applying vertical stabilization on an outdoor scene

Fig. 4.9. Two consecutive views after stabilization, using an outdoor scene

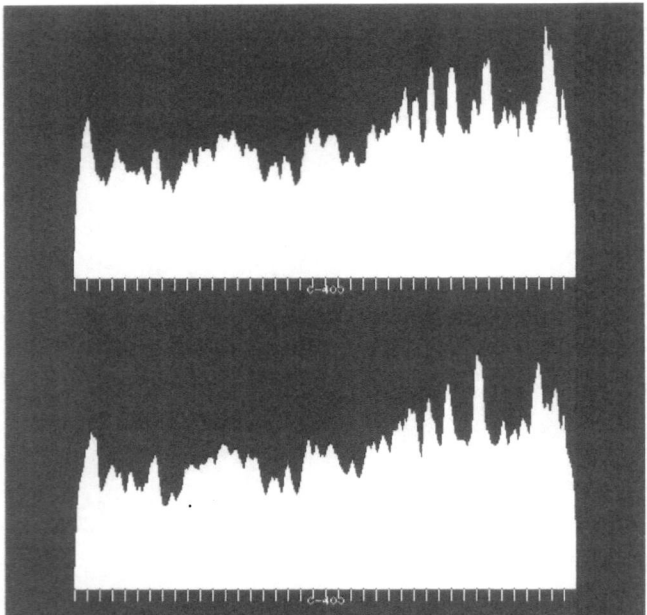

Fig. 4.10. Histogram of the horizontal coordinates for the outdoor scene

Moreover, we have observed again, all the effects expected for such an estimate: (1) if an object moves its residual disparity is visible, while stationary objects disparity is negligible (for example, the box in Fig. 4.6 is pull on the table by a non visible mechanism, and thus undergoes an additional motion which is visible on the stabilized frame); (2) remote objects have their retinal displacement canceled by global stabilization but not objects in the near (compare trees and road paintings in Fig. 4.9); (3) objects with a lot of edges (such as the telephone or trees) have a residual disparity due to uncertainty in early-vision mechanism.

Quantitatively, the average residual disparity after this stabilization is of about 0.6 pixel (standard deviation).

4.7 Conclusion

Using the vertical cue we can decompose the self-motion in different components, which can be approximated by an iterative algorithm, using reasonable assumptions about the verticality of some tokens. Then, we can correct the computed self-motion using exact equations about the structure and motion of line segments.

Moreover, performing picture rectification, we can reduce the disparity between tokens in two consecutive frames, thus improving token-tracking.

This process of rectification would also simplify the stereo geometry in the case of binocular vision.

Using the general equations of rigid motion yields problems that are ill-conditioned either when recovering self-motion or when performing structure from motion [4.20–22]. However, using the proposed method, it is still possible to obtain relevant estimates about orientation (rotation or attitude), whereas translational components are now to be estimated in a case were the rotation is negligible.

Moreover qualitative information about the 3D structure of the scene is also estimated, as done by [4.23] for instance: (1) whether or not line segments are vertical, (2) whether or not line segments or points are at horizon, (3) whether line segments are stationary with respect to the background or belong to moving objects, since the related self-motion should be biased with respect to the true self-motion.

5. Retinal Motion as a Cue for Active Vision

Most of the earlier or recent studies addressing the problem of computing structure and motion in a monocular image sequence assume that the calibration of the system is known [5.1–4], whereas we now have enough knowledge about auto-calibration, thanks to recent studies in the field [5.5–8].

In the present chapter, we address the problem of computing structure and motion of a set of planes, in a monocular image sequence, when the camera is **not** calibrated. We would like to develop a three-dimensional representation of a scene, a simple one, considering the scene is piece-wise planar. Moreover, we do not require a perfect knowledge of the rigid displacement between the two images but will estimate it as in the case of motion paradigms. What makes such motion paradigms easy to use is that we can assume the disparity between two frames to be small, leading to easy solutions for the correspondence problem. These correspondences must however be established (token-tracking) using existing tools [5.9–11].

5.1 Definition and Notation

5.1.1 Calibration: The Camera Model

We use *the standard pinhole model* for a camera, assuming that the camera performs a perfect perspective transform with center O (the camera optical center) at a distance f_0 (the focal length) of the retinal plane.

It must be noted that the pinhole model can still be used for a zoom lens if the object-to-image distance is not considered constant [5.12]. This is the case here, since we will adapt the camera metric for different object locations.

All coordinates are related to an affine frame of reference $\mathcal{R} = (O, \boldsymbol{x}, \boldsymbol{y}, \boldsymbol{z})$ *attached to the retina*, $\boldsymbol{z}$ being aligned with the optical axis, $\boldsymbol{x}$ and $\boldsymbol{y}$ being aligned with horizontal and vertical axis in the image, respectively. The retinal plane is thus perpendicular to the optical axis Oz.

We represent a 3D point $M = \boldsymbol{OM} = (X, Y, Z)^{\mathrm{T}}$ using Euclidian coordinates. Points on the retina, with horizontal and vertical pixel coordinates (u, v), will be represented as homogeneous 3D vectors: $s\,m = s\,\boldsymbol{Om} = (s\,u, s\,v, s)$, corresponding to lines of a given direction passing through the

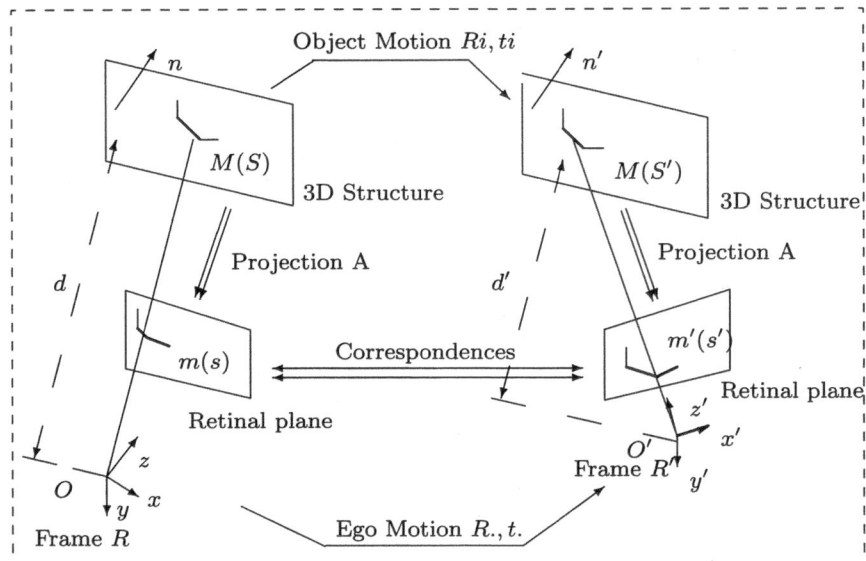

Fig. 5.1. Elements used in the definition of an H-matrix

optical center (2D projective space). Algebraic developments will thus involve
3D vectors in both cases (see Fig. 5.1).[1]

In this study, *we must not assume that the system is calibrated*. On the
contrary, we consider the following standard model for a camera, using an
A-matrix:

$$
Z\,m \;=\; \begin{vmatrix} Z\,u \\ Z\,v \\ Z \end{vmatrix}
$$

$$
= \left(\underbrace{\begin{pmatrix} \alpha_u & \gamma & u_0 \\ 0 & \alpha_v & v_0 \\ 0 & 0 & 1 \end{pmatrix}}_{A} \;\; \begin{matrix} 0 \\ 0 \\ 0 \end{matrix} \right) \cdot \begin{vmatrix} X \\ Y \\ Z \\ \mathcal{U} \in \{0,1\} \end{vmatrix}
$$

$$
= \; A \cdot M \,. \tag{5.1}
$$

For points not at infinity, we have $\mathcal{U} = 1$ and Z corresponds to the
Euclidian depth in the retinal frame of reference $\mathcal{R}$ attached to the retina
and defined previously. Equation (5.1) is still valid for "points at infinity",

---

[1] We denote vectors with **bold** letters and matrices with capital letters.

$$
\tilde{u} \cdot x = \begin{pmatrix} 0 & -u_z & u_y \\ u_z & 0 & -u_x \\ -u_y & u_x & 0 \end{pmatrix} \cdot x = u \wedge x
$$

corresponds to the cross-product, the dot-product being written as $x^{\mathrm{T}} \cdot y$. The
identity matrix is written I.

i.e. vanishing points, with $\mathcal{U} = 0$, but Z does not correspond anymore to the depth of the point. This model has already been presented in Chap. 3.

5.1.2 Motion: Discrete Rigid Displacements

We consider motions computed *in the discrete case*. We thus represent them with rigid displacements.

The 3D points of the same observed object are undergoing a rigid displacement parameterized by a rotation matrix R_i and a translation vector t_i (*object-motion*). The camera also undergoes a rigid displacement parameterized by the inverse of a rotation matrix $R_\bullet$ and the opposite of a translation vector $t_\bullet$ (*self-motion*). This last displacement corresponds to a change in the frame of reference from frame $\mathcal{R}$ at time t to frame $\mathcal{R}'$ at time $t + \Delta t$.

The rigid transformations for a point taken at time t and $t + \Delta t$ are given by the following two equations:

$$
\begin{aligned}
M'_{t+\Delta t/\mathcal{R}} &= R_i \cdot M_{t/\mathcal{R}} + t_{i/\mathcal{R}} , \\
M'_{t+\Delta t/\mathcal{R}'} &= R_\bullet \cdot M'_{t+\Delta t/\mathcal{R}} + t_{\bullet/\mathcal{R}'} \\
&= R_\bullet \cdot \left(M'_{t+\Delta t/\mathcal{R}} + t_{\bullet/\mathcal{R}} \right)
\end{aligned}
\tag{5.2}
$$

and we have the relations

$$
M'_{t+\Delta t/\mathcal{R}'} = R \cdot M_{t/\mathcal{R}} + t
\tag{5.3}
$$

$$
\text{with} \begin{cases}
R &= R_\bullet \cdot R_i = R_\bullet + \underbrace{R_\bullet \cdot (R_i - I)}_{\Delta R_i} \\
t &= t_{\bullet/\mathcal{R}'} + \underbrace{R_\bullet \cdot t_{i/\mathcal{R}}}_{\Delta t_i}
\end{cases}
$$

which can also be expressed, using the notations of (5.1):

$$
\begin{vmatrix} X' \\ Y' \\ Z' \\ \mathcal{U}' \in \{0,1\} \end{vmatrix}
= \begin{pmatrix} R & t \\ 0\ \ 0 & 1 \end{pmatrix} \cdot
\begin{vmatrix} X \\ Y \\ Z \\ \mathcal{U} \in \{0,1\} \end{vmatrix}
\tag{5.4}
$$

where $\mathcal{U} = \mathcal{U}' = 1$ for points at finite distances and $\mathcal{U} = \mathcal{U}' = 0$ for points at infinity.

Note that we have carefully made explicit the frame of reference for each quantity, since different conventions could have been used with the risk of some ambiguities.

In all the sequel, we define as stationary, an object of motion ($R_i = I, t_i = 0$).

5.1.3 Structure: Rigid Planar Structures

In order to relate 2D edges to a 3D representation we are going to consider non-parametric 2D curves or patches as tokens and assume they are related to 3D *rigid planar curves or patches.*

This means that the object lies on a plane. We parameterize this plane by a unitary normal vector $n, \|n\| = 1$, and the Euclidian distance d from the plane to the origin. The corresponding equation for a point M of this plane is

$$0 = d(M, P) = n^{\mathrm{T}} \cdot M - d \qquad \text{with} \qquad d \geq 0 . \qquad (5.5)$$

This representation is known to be regular and unambiguous [5.13]. Moreover, the quantity $|d(M, P)|$ corresponds to the Euclidian distance from a point M to the plane P, obviously null if and only if $M \in P$. Having $d(M, P) > 0$ corresponds to points behind the plane, i.e. in the half-space in the direction of n, whereas $d(M, P) < 0$ for the other half-space.

Finally d is $d(O, P)$, i.e. the distance from the plane to the origin, and d can be considered as a abbreviation of $d(O, P)$.

Considering a 2D point m and its 3D correspondent M belonging to the plane, we have obviously, from (5.1) and (5.5):

$$M = Z \, A^{-1} \cdot m \qquad (5.6)$$

with

$$Z = \frac{d}{(n^{\mathrm{T}} \cdot A^{-1} \cdot m)} \qquad (5.7)$$

which determines the 3D point location for any point in the plane, as soon as its 2D image and the 3D plane equation is given.

Finally, considering the equation of the plane in the frames $\mathcal{R}$ and $\mathcal{R}'$ of the preceding section, we easily obtain, from (5.5) and (5.3)

$$\begin{cases} n' &= R \cdot n \\ d' &= d + n^{\mathrm{T}} \cdot R^{\mathrm{T}} \cdot t \end{cases} . \qquad (5.8)$$

5.2 Using Collineations to Analyse the Retinal Motion

5.2.1 Definition

Having now a precise definition for calibration, motion and structure, we have to relate these parameters to what is measured in an image sequence: the 2D disparity between two consecutive frames.

For a 3D point M which projects in m in a frame and m' in another frame, this disparity is expressed by the relation, written from (5.1) and (5.3):

$$Z' \, m' = Z \, \underbrace{A \cdot R \cdot A^{-1}}_{Q} \cdot m + \underbrace{A \cdot t}_{s} \qquad (5.9)$$

where A is the projection matrix, (R, t) are the rigid displacement parameters (object-motion and self-motion).The quantities Q and s, so defined, will be discussed in the next section.

This equation is true for any point M undergoing a rigid displacement (R_i, t_i) between the two frames.

If we consider now M as belonging to a plane defined by (n, d), we can write from (5.9) and (5.7):

$$m' = \lambda \underbrace{\left[Q + s \cdot \left(\underbrace{A^{-1\,\mathrm{T}} \cdot \frac{n}{d}}_{\nu} \right)^{\mathrm{T}} \right]}_{H} \cdot m \qquad (5.10)$$

where $\lambda = Z/Z'$. The quantity ν, so defined, will be discussed in the next section.

This equation is true for any other point M of the plane undergoing the same rigid displacement and false for any point undergoing the same rigid displacement, but which does *not* belong to the plane.

However, it might happen that a point which does *not* belong to the plane and has *not* the same rigid displacement verifies (5.10) but this situation is not generic, considering an image sequence cannot take place during many frames. We thus will neglect this singular situation in the sequel, i.e. consider that *a point belongs to a given plane if and only if it verifies (5.10) for this plane and for all image correspondences along the sequence.*

The collineation H defined by (5.10), called a H-matrix is defined up to scale factor and is a function of 8 parameters. It is a very general computational model since it integrates correspondences, calibration, motion and structure in the same framework, as will be discussed in the next section. It provides a discrete-case model for the motion-field, which generalizes local constant or local affine motion fields [5.1]. It is an alternative for local quadratic motion fields [5.14], which only approximate the temporal relation between two frames. On the contrary, our model is exact because it accepts a set of realistic assumptions. It is a discrete model of the motion, not an approximate sampling of a continuous model.

In the case of a calibrated system (corresponding to $A = I$) this result is far from being new [5.15]. In this ideal case we have a one to one correspondence between H and $(R, \epsilon t/d, \epsilon n)$, $\epsilon = \{-1, 1\}$ [5.15]. But, in many cases, we cannot relate on calibration. We thus must introduce the A-matrix in this expression.

The underlying restriction of this model is to consider the scene as a set of piecewise planar patches undergoing rigid displacements. This corresponds to a reasonable interpretation, since (a) many objects are almost planar especially in indoors environment, (b) the perception of the object curvatures

Fig. 5.2. A few examples of images in which, clearly, the objects curvatures are not visible (e.g. the pedestrian, the walls, the road, the pieces of country-side) while each object can be locally considered as a planar patch

requires a high resolution and are often not visible even in real images (see some examples in Fig. 5.2).

Finally, it is shown in [5.16] that this model of retinal motion (represented through collineations) is only related to planar objects undergoing either a rigid-displacement or subject to an affine transformation.

5.2.2 Properties

Motion

Motion Between Two Consecutive Frames. The motion between two consecutive frames is resumed by the two quantities Q and s defined in (5.9):

The quantity Q corresponds to the *"uncalibrated rotational component of the rigid displacement"*, since

$$Q = A \cdot R \cdot A^{-1} .$$

It is not defined to a scale factor, but completely determined since $\det(Q) = 1$.

It can be interpreted as the collineation of the plane at infinity considered as undergoing the same motion as the plane corresponding to H, i.e. (R_i, t_i). Indeed, (5.4) shows that, for points at infinity, for which $\mathcal{U} = 0$, the translational component of the motion does not interfere:

$$M'_\infty = R \cdot M_\infty$$

so that (5.9) simply becomes:

$$m'_\infty = \lambda \, Q \cdot m_\infty \quad \text{with} \quad \lambda = \frac{Z}{Z'} . \qquad (5.11)$$

This last equation is also the limit of (5.10), true for any plane with the same motion, when $d \to \infty$. In other words, it corresponds to the retinal motion of remote points or "points at the horizon" undergoing the motion given by (R_i, t_i).

By definition, we consider that the plane at infinity is stationary (it is not so important since, in any case, we cannot see it!) and its collineation is then defined by

$$H_\infty = A \cdot R_\bullet \cdot A^{-1} .$$ (5.12)

So the projections (m, m') of any stationary point verify

$$Z' \, m' = Z \, H_\infty \cdot m + s_\bullet$$ (5.13)

and the H-matrix of any stationary plane verifies:

$$H = \lambda \left(H_\infty + s_\bullet \cdot \nu^{\mathrm{T}} \right)$$ (5.14)

with $s_\bullet = A \cdot t_\bullet$.

The quantity s corresponds to the *"uncalibrated translational component of the rigid displacement"*, also called *"focus of expansion"* by some scientists.

We will give two interpretations of s in special cases:

- Taking (5.13) corresponding to stationary points, that is for $R = R_\bullet$ and $t = t_\bullet$, we see that $s_\bullet$ is the epipole in the frame $\mathcal{R}'$. Indeed, the optical center of frame $\mathcal{R}$ being, by definition, stationary, it verifies (5.13). Consequently, as its projection in frame $\mathcal{R}$ is $m = (0, 0, 0)^{\mathrm{T}}$, its projection m' in frame $\mathcal{R}'$, i.e. the epipole, is $s_\bullet$.
- In the case where the two frames are simply translated one from the other ($R_\bullet = I$) and for a plane in pure translation ($R_i = I$), (5.9) becomes

$$Z' \, m' = Z \, m + s$$

and shows that the projections (m, m') of any point of such a plane is aligned with s.

H_∞ and $s_\bullet$, which correspond to the self-motion (or, equivalently to the retinal displacement of stationary objects), will play a crucial role in the sequel. The next two paragraphs are thus dedicated to their computation.

To compute the Euclidian motion, $R_\bullet$ and $t_\bullet$, from H_∞ and $s_\bullet$, we need the A-matrix, i.e. the intrinsic calibration parameters. We will see in Sect. (5.2.2) how to obtain them from H_∞.

Computing H_∞. If the self-motion is a pure translation, $H_\infty = I$. Consequently, as soon as we can detect that we are in such a situation, the estimation of H_∞ is obvious.

If not, we will make the assumption that *several points are at infinity, i.e. their retinal motion is mainly due to the rotational part of the camera-motion.*

We can, in fact, estimate the order of magnitude of the "distance of the horizon", i.e. the distance after which any translational disparity is negligible, or, in other words, the distance after which the H-matrix is mainly given by Q.

A point M at a height distance d and subject to a translation t induces an angular displacement on the retina a which can be written

$$a \simeq \frac{t0}{d} \leq \frac{||t||}{d}$$

with the notation of this small diagram:

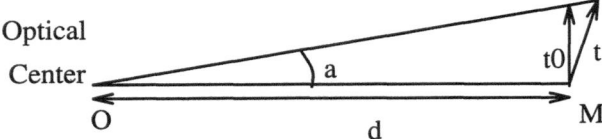

Optical Center ... O ... d ... M ... a ... t0 ... t

This angular displacement becomes negligible if a is below δ_θ the angular resolution of the visual system. We easily obtain

$$d_\infty \simeq \frac{||t||}{\delta_\theta}$$

where $||t||$ is the norm of the translation between two frames and d_∞ the distance of the horizon. For a common camera, the angular resolution, corresponding to 1 pixel, is $\delta_\theta \in [1/500, 1/1500]$ depending on the focal length. Considering a robotic application, data acquisition is 10 Hz in average, and typical velocities are 0.1 m/s to 1 m/s, thus the distance of the horizon is 10 m to 100 m. The system is then very easily "myopic". In other words it is not unrealistic to assume that there are points at the horizon in the scene.

Consequently, we can assume that *at least one estimated H-matrix of the scene corresponds to H_∞*, and we just have to choose the right one. Obviously, *we must introduce an additional knowledge about the scene to make this choice.* We can choose:

- The H-matrix corresponding to the most important number of correspondences: this is true in the case of a pure rotation, because "all stationary points are at infinity", and in the case of a small translation for all remote points.
- The H-matrix corresponding to points in the upper part of the image, because for outdoor scenes, the horizon and remote points are in this part of the image.
- The H-matrix which is compatible with the retinal motion of "vanishing points", i.e. points which correspond to the intersection of parallel lines and thus are not affected by translations but only rotations. In particular, if the vertical is detected in an image [5.17], the corresponding vanishing point can be used to eliminate H-matrices not compatible with its retinal motion.

Finally, if H_∞ has been detected at a given instant, it can be tracked from frame to frame, by simply choosing the H-matrix for which a majority of points correspond to the previous estimate.

Computing $s_{\bullet}$. Let us compute the error obtained when applying any H-matrix H of a stationary plane P, thus verifying (5.14), to any point correspondence (m, m') related to a stationary point M.

On the one hand, we have, using (5.14),

$$
\begin{aligned}
H \cdot m &= \lambda \left(H_{\infty} + s_{\bullet} \cdot \nu^{\mathrm{T}} \right) \cdot m \\
&= \lambda H_{\infty} \cdot m + \lambda \left(\nu^{\mathrm{T}} \cdot m \right) s_{\bullet}
\end{aligned}
$$

and using (5.13),

$$
\begin{aligned}
H \cdot m &= \lambda \left(\frac{Z' \, m' - s_{\bullet}}{Z} \right) + \lambda \left(\nu^{\mathrm{T}} \cdot m \right) s_{\bullet} \\
&= \lambda \frac{Z'}{Z} m' + \lambda \frac{\left[Z \left(\nu^{\mathrm{T}} \cdot m \right) - 1 \right]}{Z} s_{\bullet} \ .
\end{aligned}
$$

On the other hand, using the definition of ν and (5.6) and (5.5), we have:

$$
\begin{aligned}
Z \left(\nu^{\mathrm{T}} \cdot m \right) - 1 &= Z \left(\frac{n^{\mathrm{T}}}{d} \cdot A^{-1} \cdot m \right) - 1 = \frac{\left(n^{\mathrm{T}} \cdot M \right)}{d} - 1 \\
&= \frac{d + d(M, P)}{d} - 1 = \frac{d(M, P)}{d} \ .
\end{aligned}
$$

We thus obtain, when writing $Z = d(M, O)$ and $d = d(O, P)$:

Proposition 5.1. The error made when predicting the retinal displacement of a stationary point M using a H-matrix H corresponding to a stationary plane P is aligned with the focus of expansion and is given by the following formula:

$$
\lambda_m \, m' - H \cdot m = -\delta \, s_{\bullet} \quad ; \quad \delta = \frac{\lambda}{d(O, P)} \frac{d(M, P)}{d(M, O)} \tag{5.15}
$$

where $\lambda_m = \lambda \, Z'/Z$ is a scale factor depending on m and λ is a scale factor depending on H.

Eliminating λ_m from this last equation leads to

$$
\left[(H \cdot m) \wedge m' \right]^{\mathrm{T}} \cdot s_{\bullet} = m'^{\mathrm{T}} \cdot \tilde{s}_{\bullet} \cdot H \cdot m = 0 \tag{5.16}
$$

which allows to compute $s_{\bullet}$ up to a scale factor as soon as the projections of at least two points, not in the plane P, are given. Please note that if the point belongs to the plane, this equation is zero for any $s_{\bullet}$.

In particular, using H_{∞}, it allows us to estimate $s_{\bullet}$ considering all points in the scene and using a least-square estimation.

Detecting Non-Stationary Planes. In this paragraph, we consider that H_{∞} and $s_{\bullet}$ are known and we are interested in verifying that a given H-matrix is "compatible" with H_{∞} and $s_{\bullet}$, in the sense that it corresponds to a stationary plane.

Please note that, since we cannot observe the magnitude of the translation (see the ambiguity between structure and motion from Sect. (5.2.2)), a plane in translation t_i such that $\Delta t_i = R_\bullet \cdot t_i$ has the same direction as $t_\bullet$ (see (5.3)) will nevertheless be seen as stationary.

The question now arises whether other ambiguities can occur. Hopefully the answer is negative and we can state:

Proposition 5.2. A generic H-matrix H corresponds to a stationary plane if and only if:

$$H_\epsilon = \tilde{s}_\bullet \cdot \left(\frac{H}{||\tilde{s}_\bullet \cdot H||} - \frac{H_\infty}{||\tilde{s}_\bullet \cdot H_\infty||} \right) = 0 . \tag{5.17}$$

The only ambiguity is the scale factor ambiguity.

The proof is quite technical and not very informative. It can be found in [5.16].

As a consequence: (1) either the plane is at infinity, i.e. $\nu = 0$ which is equivalent to consider its translation as negligible, i.e. of magnitude zero; (2) or the translation of the object is zero; (3) or the translation of the background is zero; (4) or both translations equal up to a scale factor. These are not "generic" situations; they correspond to different particular cases of the scale factor ambiguity as stated in the proposition.

Equation (5.17) forms a set of 6 linear constraints for H. In practice checking the stationarity of a plane consists in comparing $||H_\epsilon||$ to a suitable threshold. This can be done using statistical estimates for H as developed in Sect. (5.3).

Motion in a Sequence. The problem of computing motion in an uncalibrated image sequence has been addressed in [5.18] but in the case of planes, the situation is very simple and given by the following proposition:

Proposition 5.3. In the case of two consecutive rigid displacements (R, t) followed by (R', t') of a plane (n, d) and with a projection matrix A, the two related H-matrices H and H' combine in a single H-matrix H'' related to the combined displacement $(R'' = R' \cdot R, t'' = R' \cdot t + t')$, i.e.

$$H'' = H' \cdot H .$$

Proof. We just have to combine

$$H'' = \underbrace{A \cdot \left(R' + \frac{t' \cdot n'^{\mathrm{T}}}{d'} \right) \cdot A^{-1}}_{H'} \underbrace{A \cdot \left(R + \frac{t \cdot n^{\mathrm{T}}}{d} \right) \cdot A^{-1}}_{H}$$

$$= A \cdot \left[R' \cdot R + R' \cdot \frac{t \cdot n^{\mathrm{T}}}{d} + \frac{t' \cdot (R^{\mathrm{T}} \cdot n)'^{\mathrm{T}}}{d'} \right.$$

$$\left. + \frac{t' \cdot n^{\mathrm{T}}}{d' d/(n'^{\mathrm{T}} \cdot t)} \right] \cdot A^{-1}$$

because using (5.3) and (5.8) and performing a few algebra yields

$$H'' = A \cdot \left\{ R' \cdot R + \left[\frac{R' \cdot t}{d} + \left(\frac{1}{d'} + \frac{1}{d\,d'/(d'-d)} \right) t' \right] \cdot n^{\mathrm{T}} \right\} \cdot A^{-1}$$

which corresponds, after a few algebra, to the claim.   □

We thus can directly combine H-matrices without making explicit the underlying Euclidian parameters. Please note that this result would have be still true if the calibration parameters were not constant. Using such combinations, we can compute the global displacement in an image sequence, given local correspondences. The recovery of 3D parameters from this global estimation is expected to be better.

Structure

Ambiguity Between Structure and Motion. The structure of the plane in a given frame, i.e. the orientation of the plane and its distance to the origin expressed in this frame, is resumed by the quantity ν defined in (5.9). It is undefined for planes intersecting the optical center, but this is not a restriction in practice because such a singularity corresponds to planes which degenerated projection is a single line.

The ambiguity between structure (ν) and motion (s) is clearly visible in (5.10). Indeed, H is a function of $s \cdot \nu^{\mathrm{T}}$ which implies that a transformation, $s \to \lambda s$ and $\nu \to \nu/\lambda$, yields the same H. It means that we will always recover ν and s, thus $t = A^{-1} \cdot s$ and $n/d = A^{\mathrm{T}} \cdot \nu$, up to a scale factor. In others words, the absolute scale of the 3D scene like the absolute distance of translation between the two frames will be unknown.

In the remainder of this section we will see how to deduce some information on the structure of the scene from H_∞, $s_\bullet$ and H-matrices of stationary planes. To deduce Euclidean information we will also need the A-matrix, i.e. the intrinsic calibration parameters. We will see in Sect. (5.2.2) how to obtain them from H_∞.

Depth Maps. Going back to (5.15), we see that we can quantify, for a stationary point which does not belong to a stationary plane of H-matrix H, the related error made when predicting its projected motion using H. More precisely, this error is related to a quantity δ called "relative pseudo-distance" since it is the ratio between the distance from M to the plane and M to the optical center, scaled by the distance from the plane to the optical center, up to a scale factor. From (5.15), we have:

$$H \cdot m \wedge m' = \delta \left[s_\bullet \wedge m' \right] \tag{5.18}$$

which leads to

$$\delta = \frac{\| H \cdot m \wedge m' \|}{\| s_\bullet \wedge m' \|} \tag{5.19}$$

and allows, knowing $s_\bullet$ and H, to determine δ for each point not in the plane.

The vectorial equation (5.18) of cross-products gives us two scalar equations to estimate δ, thus, in practice, one equation too many. This means that, if the retinal displacement $(m' - m)$ is only known in one direction, as it appears along curves for which only the normal displacement can be recovered, one can, in general, still evaluate δ. If the retinal displacement is completely known, the estimation of δ can, then, be weighted considering some knowledge about the 2D correspondence.

It is straightforward to verify that (5.15) is still valid if $H = H_\infty$ but with $\delta = 1/Z$ and $\lambda_m = Z'/Z$. Knowing H_∞ and $s_\bullet$, we can thus directly compute *the depth map in the retinal frame of reference* without explicitly recovering the matrix A and the Euclidian motion. In other words, we only reconstruct the Euclidian structure up to a collineation of the image, given by the matrix A, which is an *affine transformation* of the plane. To completely reconstruct the Euclidian structure, we need the A-matrix.

Plane Structure. Knowing H_∞ and $s_\bullet$ we can recover the uncalibrated structure ν of a stationary plane of H-matrix H.

Indeed, extracting λ from (5.14), we obtain

$$||\tilde{s}_\bullet \cdot H|| = \lambda ||\tilde{s}_\bullet \cdot H_\infty|| \, ,$$

which substituted in (5.14) gives:

$$\nu = \frac{||\tilde{s}_\bullet \cdot H_\infty||}{||s_\bullet||} \left(\frac{H}{||\tilde{s}_\bullet \cdot H||} - \frac{H_\infty}{||\tilde{s}_\bullet \cdot H_\infty||} \right)^{\mathrm{T}} \cdot \frac{s_\bullet}{||s_\bullet||} \, . \tag{5.20}$$

We can also calculate the evolution of this parameter, using its definition given by (5.10) and using (5.8):

$$\nu = \frac{A^{-1 \, \mathrm{T}} \cdot R^{\mathrm{T}} \cdot n'}{d' - n'^{\mathrm{T}} \cdot t} = \frac{d' A^{-1 \, \mathrm{T}} \cdot R^{\mathrm{T}} \cdot A^{\mathrm{T}} \cdot \nu'}{d' - d' \nu'^{\mathrm{T}} \cdot A \cdot t}$$

which leads to:

$$\nu = \frac{H_\infty^{\mathrm{T}} \cdot \nu'}{1 - s_\bullet^{\mathrm{T}} \cdot \nu'} \, . \tag{5.21}$$

This last equation allows us to relate the parameters corresponding to the structure of the plane between two frames without making explicit the Euclidian parameters. This equation is to be used in a feedforward scheme, to predict the next estimation of ν.

To recover the Euclidian parameters of the plane, we need the A-matrix. Then, we have, from (5.10) and (5.9):

$$n = \frac{A^{\mathrm{T}} \cdot \nu}{||A^{\mathrm{T}} \cdot \nu||} \quad ; \quad \frac{d}{||t||} = \frac{1}{||A^{\mathrm{T}} \cdot \nu|| \, ||A^{-1} \cdot s_\bullet||} \, . \tag{5.22}$$

Being obtained from a monocular image sequence the distance d from the plane to the origin is scaled by the inverse of the magnitude of the translation. This corresponds to a "time to collision", i.e. the time, considering a constant velocity of amplitude $v = ||\boldsymbol{t}||/\Delta_T$, for the plane to intersect the optical center of the camera. This quantity cannot be recovered without the calibration parameters.

Calibration

Weak Calibration: Relation with the Fundamental Matrix. In the uncalibrated case, the constraint between points in correspondence is entirely represented by a special matrix, the fundamental matrix F. This matrix is also called the "essential matrix in the uncalibrated case" considering the original Longuet–Higgins equation [5.19], now generalized. It turns out that there is a deep relationship between the H-matrix of any stationary plane and F defined by [5.5, 20]. This relationship has been already derived in Sect. (5.2.2), since (5.16) implies that

$$F = \tilde{s}_{\bullet} \cdot H \ . \tag{5.23}$$

This equation tells us that $H^{\mathrm{T}} \cdot F$ is an anti-symmetrical matrix which can be expressed:

$$F^{\mathrm{T}} \cdot H + H^{\mathrm{T}} \cdot F = 0 \ . \tag{5.24}$$

This last equation yields 5 equations which allows, us either to constrain, knowing F, the value of the H-matrix, as in [5.21] for a stereo rig, or conversely, to estimate F, knowing at least two H matrices as in [5.22].

Let us discuss the precision of the two approaches in the following framework: real cameras with dynamic parameters (approximate calibration) and real-time image sequences (small displacements), thus a very common situation in active vision.

The approximate calibration is expressed by the following equation:

$$A = A_0 \cdot (I + \delta A) \tag{5.25}$$

where A_0 is known and δA is to be determined, but is expected to be close to zero.

The hypothesis of small displacements lets us linearize the rotation matrix. Thus, if we represent the rotations using the exponential of an anti-symmetrical matrix, that is

$$R = e^{\tilde{r}} \tag{5.26}$$

where $\boldsymbol{r}$ is a vector aligned with the rotation axis and which magnitude is equal to the angle of the rotation [5.13], we can compute the expansion of the H-matrix and obtain, after a straightforward derivation:

$$A_0^{-1} \cdot H \cdot A_0 = \lambda \left[I + \tilde{r} + \frac{1}{2}\tilde{r}^2 + \frac{\boldsymbol{t} \cdot \boldsymbol{n}^{\mathrm{T}}}{d} + (\delta A \cdot \tilde{r} - \tilde{r} \cdot \delta A) \right] + o(\epsilon) \tag{5.27}$$

with

$$\epsilon = \left(||r||^3 + ||r||^2\,||\delta A|| + ||r||\,||\delta A||^2 + ||\tfrac{t}{d}||\,||\delta A|| \right)\ .$$

Similarly, we have

$$
\begin{aligned}
A_0^{-1} \cdot F \cdot A_0 &= \lambda \left[\tilde{t} + \tilde{t} \cdot \tilde{r} + \left(\delta A \cdot \tilde{t} - \tilde{t} \cdot \delta A \right) \right] \\
&\quad + o \left[\left(||\delta A||^2 + ||\tilde{r}||^2 \right) ||t|| \right]\ .
\end{aligned}
\tag{5.28}
$$

As a consequence, the calibration parameters are, up to a first order, related to the translational motion for F, and the rotational motion for H, the former being known to be less stable than the latter [5.23, 24]. In addition, for small translation F is ill-conditioned, whereas H is not. Moreover, for 3D points with a very large depth, their contribution to the estimate of F vanishes, and F is ill-conditioned again.

Then, although the F-matrix has the advantage of being independent of the scene structure, to be defined even if the tokens are not coplanar, and to provide a direct solution of calibration parameters, it is numerically less stable. It is the price to pay when the formalism is very general.

Therefore, considering (5.24), we better use it to compute F knowing H. This is the approach followed by [5.22] in which it is demonstrated that using (5.24) is an efficient way to estimate the F-matrix and thus perform auto-calibration. In this book, we complete this work by proposing a robust estimation of the H-matrix.

Strong Calibration: Relation with the Intrinsic Parameters. Let us now describe how to compute the A-matrix from H_∞.

If we write that $R_\bullet = A^{-1} \cdot H_\infty \cdot A$ is an orthogonal matrix we obtain

$$
\begin{aligned}
R_\bullet \cdot R_\bullet^{\mathrm{T}} &= I\ , \\
A^{-1} \cdot H_\infty \cdot A \cdot A^{\mathrm{T}} \cdot H_\infty^{\mathrm{T}} \cdot A^{-1\,\mathrm{T}} &= I\ , \\
H_\infty \cdot \underbrace{A \cdot A^{\mathrm{T}}}_{K} \cdot H_\infty^{\mathrm{T}} &= \underbrace{A \cdot A^{\mathrm{T}}}_{K}\ .
\end{aligned}
$$

We thus have

$$H_\infty \cdot K \cdot H_\infty^{\mathrm{T}} = K \tag{5.29}$$

where

$$
K = A \cdot A^{\mathrm{T}} = \begin{pmatrix}
\underbrace{\alpha_u^2 + \gamma^2 + u_0^2}_{b_u} & \underbrace{u_0\,v_0 + \gamma\,\alpha_v}_{b_c} & u_0 \\
\underbrace{u_0\,v_0 + \gamma\,\alpha_v}_{b_c} & \underbrace{\alpha_v^2 + v_0^2}_{b_v} & v_0 \\
u_0 & v_0 & 1
\end{pmatrix}
\tag{5.30}
$$

is a symmetric positive matrix in one-to-one correspondence with the A-matrix since:

$$\alpha_v = \sqrt{b_v - v_0^2}\ ;\quad \gamma = \frac{b_c - u_0\,v_0}{\alpha_v}\ ;\quad \alpha_u = \sqrt{b_u - u_0^2 - \gamma^2} \tag{5.31}$$

is unambiguous since $\alpha_u > 0$ and $\alpha_v > 0$.

K defines, in fact, the retinal image of the absolute conic at infinity which equation is $0 = m^{\mathrm{T}} \cdot K^{-1} \cdot m$, that is the angular relations between each optical ray, or in other words, the intrinsic calibration or the Euclidian geometry of the system [5.5].

We notice that K depends linearly on H_∞. Let us analyse how many equations are generated. Because it involves an equality between 3×3 symmetric matrices, we have only 6 independent equations, up to a scale factor, thus only 5, but these equations are constrained by the fact that we must have $\det(H_\infty \cdot K \cdot H_\infty^{\mathrm{T}}) = \det(K)$, always verified since $\det(H_\infty) = 1$. We thus have one equation always verified, and (5.29) provides finally *4 independent equations on the intrinsic parameters, linear with respect to vector of unknown* $(u_0, v_0, b_u, b_v, b_c)$. This has been also demonstrated in [5.18] using the following decomposition:

Proposition 5.4. *(Luong Decomposition of H_∞)* If we parametrize the rotation matrix with $R = e^{\tilde{r}}$, as in (5.26) we have:

$$\log(H_\infty) = K \cdot \tilde{\rho} \quad \text{with} \quad \rho = \frac{A \cdot r}{\det(A)} \ .$$

As a consequence, for a given rotation, K is not measurable in the direction of ρ.

Proof. The decomposition is straightforward to derive:[2]

$$
\begin{aligned}
H_\infty &= A \cdot e^{\tilde{r}} \cdot A^{-1} = e^{A \cdot \tilde{r} \cdot A^{-1}} = e^{A \cdot (A^{-1} \cdot A \cdot r) \cdot A^{-1}} \\
&= e^{A \cdot A^{\mathrm{T}} \, \det(A^{-1}) (\tilde{A} \cdot r)} = e^{K \cdot \tilde{\rho}} \ .
\end{aligned}
$$

$\square$

As a consequence, we can recover, in theory, the intrinsic calibration parameters as soon as two displacements (i.e. three views) are given, with two rotations not performed around the same axis. We, in that case, have an overdetermined linear system, and we can compute the intrinsic calibration parameters using a linear algorithm. This is not the case for other approaches [5.5, 20]. Finally, in practice, when using three views, that is two displacements, we better use one inducing horizontal disparity and one inducing vertical disparity to obtain a good estimate, as discussed in [5.25].

Let us discuss in more detail, why we can analyse the intrinsic parameters using the collineation of the plane at infinity in an apparently simpler way, than using the fundamental matrix. As explained before, the matrix H_∞ corresponds to the collineation of the plane at infinity. But knowing the plane at infinity, means knowing the *affine geometry* of the scene [5.13]. In other

---

[2] We make use of the following relation:
$$(M \cdot x) \wedge (M \cdot y) = \underbrace{\det(M) \, M^{-1\,\mathrm{T}}}_{M^*} \cdot (x \wedge y)$$
or equivalently
$$(\tilde{M \cdot x}) \cdot M = \det(M) \, M^{-1\,\mathrm{T}} \cdot \tilde{x} \ .$$

words, *the use of the matrix H_∞ provides an intermediate representation of the motion in which the affine geometry of the scene has been made explicit.* A projective representation would have involved F-matrices only. An Euclidian representation would have involved rotation matrices and translation vectors.

Moreover, we have seen that it is possible, for a rather large class of real scenes, to infer the collineation of the plane at infinity, especially when considering points "at the horizon", that is with a negligible depth, as for a similar paradigm in [5.17].

5.3 Estimation of Retinal Motion from Correspondences

5.3.1 Estimation of a H-Matrix from Point Correspondences

Let us now turn to the problem of estimating a H-matrix This estimation is not straightforward for several reasons: ill-conditioning, singularities, aperture problem, occlusions, and objects in motion.

In order to deal with all these problems we first tackle with a simple case considering point to point correspondences between two sets of coplanar points. We will see at the end of this section, that the generalization to other primitives is straightforward.

Modeling the Uncertainty on Point Correspondences. Let us consider a set of points $\{m_i\}_{i=1...N}$ in one frame and their correspondents $\{m_i'\}_{i=1...N}$ in the other frame.

In order to introduce some robustness in our estimation, we require that each correspondence is given with a covariance Λ_i, as discussed now.

The main source of error, considering such point correspondence, is related to the uncertainty in the pixel localization and the uncertainty in the correspondence, the latter being more important than the former. A simple way to represent this uncertainty is to consider for each point of correspondence that the value of m_i is given with a certain uncertainty corresponding to a covariance Λ_{m_i}. The uncertainty is "thus carried by m_i", while m_i' is taken as a deterministic value, neglecting its localization error in the image, i.e. $\Lambda_{m_i'} = 0$ and $\Lambda_{m_i} = \Lambda_i$. Such a model corresponds, for instance, to the uncertainty of a correlation operator as discussed in the next section. If we do not have any particular knowledge on the nature of this uncertainty, we consider the statistical distribution as being isotropic, and each measure to be independent from the others. It allows us to make, for some situation, use of the following simplifying assumptions: $\Lambda_{m_i} = \sigma^2 \cdot I$.

Let us note

$$C_i = \begin{pmatrix} I_{uu} & I_{uv} \\ I_{uv} & I_{vv} \end{pmatrix}$$

the 2D covariance attached to the 2D retinal correspondence. We then have, with the notation $m_i = (u_i, v_i, 1)$ a covariance $\Lambda_{m_i} = \Lambda_i$ of the form[3]

$$\Lambda_i^{-1} = \begin{pmatrix} \dfrac{\begin{pmatrix} I_{vv} & -I_{uv} \\ -I_{uv} & I_{uu} \end{pmatrix}}{I_{uu}\,I_{vv} - I_{uv}^2} & \begin{matrix} 0 \\ 0 \end{matrix} \\ \begin{matrix} 0 \qquad 0 \end{matrix} & 0 \end{pmatrix} . \tag{5.32}$$

The matrix Λ_i^{-1} is the inverse of the covariance matrix, also called *information matrix*, related to the measure obtained from the punctual correspondence between m_i and m_i'.

Looking for a Robust Criterion. The fact that H maps $\{m_i\}_{i=1...N}$ on $\{m_i'\}_{i=1...N}$ can be written $\forall i$, $\lambda_i m_i' = H \cdot m_i$ for some λ_i.

Let us first introduce some notations which will allow us to write our measurement equations using homogeneous quantities and lead to much simpler formulation in the sequel.

In the image, local correspondences are given between two retinal points (u_i, v_i) and (u_i', v_i'). More precisely, if we use the following notation for the components of a 3×3 H-matrix:

$$H = \begin{pmatrix} \underbrace{H_{00}\ H_{01}\ H_{02}}_{h_0^{\mathrm{T}}} \\ \underbrace{H_{10}\ H_{11}\ H_{12}}_{h_1^{\mathrm{T}}} \\ \underbrace{H_{20}\ H_{21}\ H_{22}}_{h_2^{\mathrm{T}}} \end{pmatrix}$$

the retinal coordinates $(\bar{u}_i', \bar{v}_i')$ of the image of a 2D point written $m_i = (u_i, v_i, 1)$ though the collineation given by H and the related retinal error ϵ_i are

$$\begin{cases} \bar{u}_i' &= \dfrac{h_0^{\mathrm{T}} \cdot m_i}{h_2^{\mathrm{T}} \cdot m_i} \\ \bar{v}_i' &= \dfrac{h_1^{\mathrm{T}} \cdot m_i}{h_2^{\mathrm{T}} \cdot m_i} \end{cases}$$

and

$$\epsilon_i = \left| \begin{matrix} \zeta_i \\ 0 \end{matrix} \right. = \frac{1}{h_2^{\mathrm{T}} \cdot m_i} \, (\lambda_i\, m_i' - H \cdot m_i)$$

with

---

[3] We make use of the following property:

$$y = f(x) \quad \Rightarrow \quad \Lambda_x^{-1} \simeq \frac{\partial f(x)}{\partial x}^{\mathrm{T}} \cdot \Lambda_y^{-1} \cdot \frac{\partial f(x)}{\partial x} .$$

See for instance [5.26] for details.

$$\zeta_i = \left| \begin{array}{c} u_i' - \bar{u}_i' \\ v_i' - \bar{v}_i' \end{array} \right. \qquad \text{and} \qquad \lambda_i = \boldsymbol{h}_2^{\mathrm{T}} \cdot m_i \; .$$

When combining several measures a natural requirement is to minimize the average mean-square retinal error, for each correspondence, which is known to be an efficient criterion [5.9, 27, 28]. This situation can be interpreted as follows: if there is no disparity between two frames, there is no information about the projected motion, and no possibility to analyze the visual scene. Conversely, providing a model of motion is available, one may cancel all the disparity between two consecutive frames by a suitable "reprojection". This means that we have extracted all the information about the projected motion. In other words, an image sequence is perfectly stabilized if and only if motion knowledge allows us to cancel the disparity between two frames and no ambiguity occurs. In the presence of uncertainties, because we have implicitly considered that the retinal location is subject to a random additive perturbation, the optimal strategy is to minimize the Mahalanobis distance (pseudo-chi-square criterion) [5.29], introducing some robustness with respect to noisy data.

This yields to the following least-square criterion:

$$\begin{aligned} \mathcal{J}_1 &= \sum_i \left(\zeta_i^{\mathrm{T}} \cdot C_i^{-1} \cdot \zeta_i \right) \\ &= \sum_i \frac{1}{\boldsymbol{h}_2^{\mathrm{T}} m_i} (\lambda_i m_i' - H \cdot m_i)^{\mathrm{T}} \cdot \Lambda_i^{-1} \cdot (\lambda_i m_i' - H \cdot m_i) \end{aligned}$$

which we will now modify to lead to a more computable and robust criterion.

Firstly, such a criterion is non-quadratic with respect to the motion parameters ($\boldsymbol{h}_2$ is in the denominator in our case), while we would like to avoid nonlinear minimization for fast implementation. A simple strategy is to weight the criterion, by a fixed quantity using a priori estimation of H, which will be denoted by $\hat{\boldsymbol{h}}_2$. As soon as this quantity is fixed, we obtain a quadratic and positive criterion in H and $\{\lambda_i\}_{i=1...N}$ and the related equations vanish if and only if the related criterion reaches its minimum. In the presence of noise, the related equations are not completely verified, but the minimization is expected to provide a robust estimate. This sub-optimal but fast strategy corresponds asymptotically to the expected criterion [5.30].

Secondly, we would like to weight each measure considering other heuristics to be able to adapt the criterion to various other metrics. This is the role of κ_i in the new formulation of the criterion:

$$\begin{aligned} \mathcal{J}_1' &= \sum_i \kappa_i \left(\zeta_i^{\mathrm{T}} \cdot C_i^{-1} \cdot \zeta_i \right) \\ &= \sum_i \underbrace{\frac{\kappa_i}{\hat{\boldsymbol{h}}_2^{\mathrm{T}} m_i}}_{\mu_i} (\lambda_i m_i' - H \cdot m_i)^{\mathrm{T}} \cdot \Lambda_i^{-1} \cdot (\lambda_i m_i' - H \cdot m_i) \; . \end{aligned}$$

In the following derivations and without loss of generality we are going to integrate μ_i in the information matrix and write the equations as if $\mu_i = 1$, while $\Lambda_i^{-1} \to \mu_i\, \Lambda_i^{-1}$.

An Example of Choice of κ_i. When we analyse the motion of a planar patch, we would like the system to analyse connected planar structures, and consider that structures with some "compactness", belong to the same planar patch.

However it might occur that some points taken into account are far from the expected structure. for instance, it appears that some points (they are on a line!) belong to the intersection of two different planes. We must decide which plane they belong to. As an example, in Fig. 5.3 the crosse motion is compatible with both the motion of plane of the table and the motion of the frontal face of the box. However, it is natural to consider these crosses as being on the table, because *in this image, neighbor points of the crosses are on table.* This proximity constraint must be integrated in the procedure.

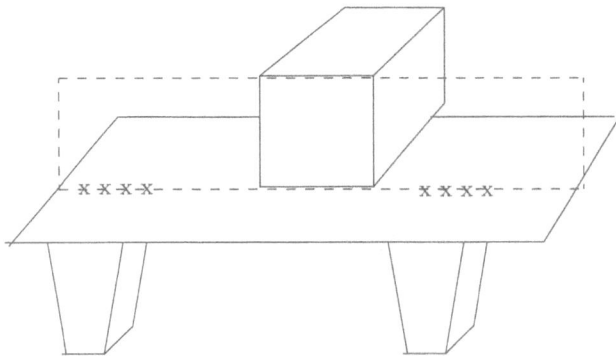

Fig. 5.3. Ambiguity in the segmentation, the case of a box on a table

In order to fit with this requirement, we have defined

$$\kappa_i = \frac{1}{1 + \left[\dfrac{d(m_i, g)}{d_0}\right]^2}$$

where g is the center of gravity of the set of points which correspond with which the collineation is to be estimated. The parameter d_0 corresponds to a distance after which the contribution of each point to the criterion is divided by at least 2. By choosing d_0 equal to twice the spatial variance of the point locations, we are going to reduce the contribution of about 10 % of outliers, which seems to be a reasonable choice. Please note that g and d_0 are estimated using *a priori* information on the collineation and the points related to this collineation. During the bootstrapping phase, we will simply take $\kappa_i = 1$.

Computation of the Optimal Estimate. It is then straightforward to compute the set of normal equations[4]

$$
\begin{cases}
0 = \dfrac{1}{2}\dfrac{\partial \mathcal{J}_1}{\partial H}^{\mathrm{T}} = -\sum_i \Lambda_i^{-1} \cdot (\lambda_i m'_i - Hm_i) \cdot m_i^{\mathrm{T}}, \\
0 = \dfrac{1}{2}\dfrac{\partial \mathcal{J}_1}{\partial \lambda_i} = (\lambda_i m'_i - Hm_i)^{\mathrm{T}} \cdot \Lambda_i^{-1} \cdot m'_i
\end{cases}
$$

$$
\Rightarrow \quad 0 = \sum_i m'_i \cdot \underbrace{\frac{(m'^{\mathrm{T}}_i \cdot \Lambda_i^{-1} \cdot H \cdot m_i)}{(m'^{\mathrm{T}}_i \cdot \Lambda_i^{-1} \cdot m'_i)}}_{\lambda_i} \cdot m_i^{\mathrm{T}} - H \cdot m_i \cdot m_i^{\mathrm{T}} \quad (5.33)
$$

which are linear equations for H. Note that this equation is normalized with respect to the magnitude of the homogeneous quantities: if we multiply m'_i by a given quantity, say k' and/or m_i by another quantity k, therefore Λ_i is multiplied by k^2, these two factors vanish in (5.33).

For some derivations, because we are in a situation of small displacements $H = I + \delta_H$ we have

$$
\|H \cdot m_i\| = \|m_i\| \left[1 + o\left(\left(\frac{\|\delta_H \cdot m_i\|}{\|m_i\|} \right)^2 \right) \right] \simeq 1 \quad (5.34)
$$

up to the second order, and from (5.33) we observe, for small displacements that

$$
\lambda_i \|m'_i\| = \|H \cdot m_i\| \quad \Rightarrow \quad \lambda_i \simeq \frac{\|m_i\|}{\|m'_i\|} \simeq 1. \quad (5.35)
$$

This approximation will be used in the sequel.

This can be summarized as follows:

Proposition 5.5. The H-matrix which maps $\{m_i\}_{i=1...N}$ onto $\{m'_i\}_{i=1...N}$ verifies the following 3×3 linear equations:

$$
\begin{aligned}
0_{3\times 3} &= f(H, \{m_i\}, \{m'_i\}) \\
&= \sum_i \Lambda_i^{-1} \cdot \left(m'_i \cdot \frac{m'^{\mathrm{T}}_i \cdot \Lambda_i^{-1} \cdot H \cdot m_i}{m'^{\mathrm{T}}_i \cdot \Lambda_i^{-1} \cdot m'_i} \cdot m_i^{\mathrm{T}} \right)
\end{aligned}
$$

---

[4] In these derivations we make use of the following algebraic derivation rules for $J = \|b - A \cdot c\|^2$:

$$
\frac{1}{2}\frac{\partial J}{\partial b}^{\mathrm{T}} = b - A \cdot c,
$$

$$
\frac{1}{2}\frac{\partial J}{\partial c}^{\mathrm{T}} = A^{\mathrm{T}} \cdot (A \cdot c - b),
$$

$$
\frac{1}{2}\frac{\partial J}{\partial A}^{\mathrm{T}} = (A \cdot c - b) \cdot c^{\mathrm{T}},
$$

as easily computable.

$$-H \cdot m_i \cdot m_i^{\mathrm{T}} \Bigg) . \tag{5.36}$$

The obtained solution is defined up to a scale factor, and is optimal in the presence of a Gaussian additive noise.

Three technical points arise here:

- The H-matrix is only defined up to a scale factor and these equations are not linear, but homogeneous, and the optimal solution must be found solving an eigenvalue problem.
- Depending on the application, it is not always suitable to use a collineation to represent the retinal displacement. On the contrary, it has been shown that a simpler model such as an affine transformation is sufficient to segment objects in a scene [5.1], and in some particular cases (panoramic motion) a model of a constant visual field might be enough to detect and track an object in motion [5.31].
- Depending on the number of correspondences available, (5.36) might not have a unique solution (up to a scale factor).

We must thus design a mechanism which avoids this scale factor indetermination, does not use a huge set of parameters when unnecessary and allows to compute an estimate of the H-matrix even if not many correspondences are given. In order to integrate these requirements, we have implemented the estimation of an H-matrix under the following possible sets of linear constraints, as shown in Table 5.1.

The original criterion $\mathcal{J}_1'$ is thus modified in order to integrate linear constraints, as developed in [5.26]. It has been demonstrated that minimizing a quadratic criterion under a set of linear constraints is equivalent to working in the affine subspace defined by these constraints. Therefore we do not need to derive again all expressions but have only to integrate these constraints in the estimate. Viéville, Sander [5.26] and Faugeras [5.13] provide a detailed description of such an implementation. In any case we avoid the scale factor indetermination by setting $H_{22} = 1$ which is a reasonable choice, since, in an image sequence, the H-matrix is not far from the identity matrix.

The different choices of H-matrices correspond to plausible retinal motions in an active visual system as discussed by [5.1], such as panoramic rotations of a camera. We also have integrated the possibility that the system is stabilized with respect to the vertical in the image as developed in [5.17]. This aspect will be discussed in the next part of this book.

As a consequence, we are not going to estimate a single H-matrix, but several concurrent H-matrices and choose the most relevant estimate. We expect from this mechanism, that the H-matrix will be estimated with a minimum number of degrees of freedom d_H. The next problem is then to decide how to choose between these different models. This is going to be developed in the next section.

Table 5.1. Using several representations for an H-matrix

Type of motion	H-matrix	Constraints	Number of degrees of freedom
Constant motion	$\begin{pmatrix} 1 & 0 & a \\ 0 & 1 & b \\ 0 & 0 & 1 \end{pmatrix}$	$\begin{cases} H_{00} = 1 \\ H_{01} = 0 \\ H_{10} = 0 \\ H_{11} = 1 \\ H_{20} = 0 \\ H_{21} = 0 \\ H_{22} = 1 \end{cases}$	$d_H = 2$
Horizontal-affine motion	$\begin{pmatrix} c & 0 & a \\ 0 & 1 & b \\ 0 & 0 & 1 \end{pmatrix}$	$\begin{cases} H_{01} = 0 \\ H_{10} = 0 \\ H_{11} = 1 \\ H_{20} = 0 \\ H_{21} = 0 \\ H_{22} = 1 \end{cases}$	$d_H = 3$
Vertical-affine motion	$\begin{pmatrix} 1 & 0 & a \\ 0 & d & b \\ 0 & 0 & 1 \end{pmatrix}$	$\begin{cases} H_{00} = 1 \\ H_{01} = 0 \\ H_{10} = 0 \\ H_{20} = 0 \\ H_{21} = 0 \\ H_{22} = 1 \end{cases}$	$d_H = 3$
1D affine motion	$\begin{pmatrix} c & 0 & a \\ 0 & d & b \\ 0 & 0 & 1 \end{pmatrix}$	$\begin{cases} H_{01} = 0 \\ H_{10} = 0 \\ H_{20} = 0 \\ H_{21} = 0 \\ H_{22} = 1 \end{cases}$	$d_H = 4$
Affine motion	$\begin{pmatrix} c & e & a \\ f & d & b \\ 0 & 0 & 1 \end{pmatrix}$	$\begin{cases} H_{20} = 0 \\ H_{21} = 0 \\ H_{22} = 1 \end{cases}$	$d_H = 6$
Projective motion	$\begin{pmatrix} c & e & a \\ f & d & b \\ g & h & 1 \end{pmatrix}$	$\begin{cases} H_{22} = 1 \end{cases}$	$d_H = 8$

5.3.2 Performing Tests on the Estimate

Computation of Covariances. Let us first compute the covariance of the H-matrix, considering the previous estimation.

We consider the case of small displacements, i.e. H is not far from the identity matrix. From (5.35), we have $\lambda_i \simeq 1$. Therefore, the 3×3 components of (5.36) are

$$0 = f^{ab} = \sum_i \sum_h (\Lambda_i^{-1})_h^a \left(m_i'^h \, m_i^b - \sum_k H_k^h m_i^k \, m_i^b \right)$$

and obviously

$$\frac{\partial f^{ab}}{\partial H_k^h} = \sum_i (\Lambda_i^{-1})_h^a \, m_i^k \, m_i^b \quad , \quad \frac{\partial f^{ab}}{\partial m_i'^k} = (\Lambda_i^{-1})_k^a \, m_i^b \, .$$

Considering a first order expansion for the computation of covariances and independency between $\{m_i'\}$, we have

$$\Lambda_{f^{ab}}^{cd} = \sum_i \sum_{jk} \frac{\partial f^{ab}}{\partial m_i'^j} \cdot \Lambda_i^{jk} \cdot \frac{\partial f^{cd}}{\partial m_i'^k}^{\mathrm{T}} = \sum_i (\Lambda_i^{-1})_h^a \, m_i^k \, m_i^d \, .$$

Using the implicit-function theorem applied to the computation of covariances as given for instance in [5.29], we can write the following first order expansion:

$$\Lambda_H^{cd\,-1} = \frac{\partial f^{cd}}{\partial H}^{\mathrm{T}} \cdot \Lambda_{f^{cd}}^{-1} \cdot \frac{\partial f^{cd}}{\partial H} \, .$$

This simplifies to

$$\Lambda_H^{cd\,-1} = \frac{\partial f^{cd}}{\partial H}^{\mathrm{T}}$$

due to the previous calculation. This finally leads to the following formula:

Proposition 5.6. The covariance of the estimate of H, for H-matrices not far from the identity matrix, using first order expansions, can be written

$$\left(\Lambda_{H_a^b}^{cd} \right)^{-1} \simeq \underbrace{\left[\sum_i \sum_h (\Lambda_i^{-1})_h^a \cdot m_i^h \, m_i^d \right]}_{I_H^{ad}} \delta_{cb} \, .$$

This covariance allows us to estimate the precision of the estimate of the H-matrix, based on (5.36). The proposed approximation is a compromise between the accuracy of the formula and its simplicity. In our solutions, the matrix I_H^{ad} collects all the information obtained from the measurements.

Since H is defined up to a scale factor, the uncertainty must be normally represented using directional statistics as, for instance, the Bingham distribution (see [5.32] for a review on directional statistics). Such a representation is often used in vision, for quantities such as vanishing points [5.33]. However, we do not have this problem here, since we have used the constraint $H_{22} = 1$. Therefore, all H-matrices are represented by a 8-dimensional vector, in a Euclidian space. Formally: we have not represented the subset of H-matrices we are dealing with and are in the neighborhood of the identity matrix on a unit hyper-sphere but on a hyper-plane. As a consequence the covariance matrix of the 8-dimensional vector is given by the restriction of the previous covariance to the given hyper-plane, as developed in [5.26].

As a consequence, we can take into account an initial estimate H_0 of the H-matrix, with an associated covariance

$$\left[\Lambda^{c\,d}_{(H_0)^b_a} \right]^{-1} = I^{a\,d}_{H_0} \delta_{b\,c} \ .$$

Note that the chosen covariance has the form given in the last proposition. We therefore can combine this estimate with a set of correspondences using the following criterion, related to a distance of Mahalanobis:

$$
\begin{aligned}
\mathcal{J}''_1 &= \sum_{abcd} [H^a_c - (H_0)^a_c]^{\mathrm{T}} \left[\Lambda^{c\,d}_{(H_0)^b_a} \right]^{-1} [H^b_d - (H_0)^b_d] + \mathcal{J}'_1 \\
&= \mathrm{trace} \left[(H - H_0)^{\mathrm{T}} \cdot I_{H_0} \cdot (H - H_0) \right] + \mathcal{J}'_1
\end{aligned}
$$

which generalizes the criterion used in Prop. 5.5 and yields the following normal equation:

$$0_{3\times3} = I_{H_0} \cdot (H - H_0) + f\left(H, \{m_i\}, \{m'_i\}\right) \ .$$

This is to be used when computing a H-matrix in an image sequence, the previous estimate of the H-matrix being used in the process. Our framework thus corresponds to a linear Kalman-filter of the H-matrix, given a set of observation and an initial estimate.

Statistical Inference. Following this formalism and using well known tools [5.29], we can analyse the residual errors obtained from these estimates. They will allow us to perform some very useful statistical tests, as developed now:

$\mathcal{T}_1$: *Is a point correspondence not compatible with a given collineation?* Let us assume that we have for a point correspondence $\lambda\, m'_i = H \cdot m_i + \epsilon_i$ where $bfz\epsilon_i$ is expected to be a white Gaussian noise, with zero mean and a covariance Λ_i. The H-matrix has been estimated with a information matrix I_H. The information matrix, combining uncertainty due to the estimation of H and the point correspondence is

$$\Lambda_{\epsilon_i} = \Lambda_i + \frac{1}{m_i^{\mathrm{T}} \cdot I_H \cdot m_i} I \ .$$

This can be easily obtained considering first order expansions. Now, if this model is true the quantity

$$\Xi^2(m_i, m_i', \Lambda_i^{-1}, H, I_H) = (\lambda_i\, m_i' - H \cdot m_i)^{\mathrm{T}} \cdot \Lambda_{\epsilon_i}^{-1} \cdot (\lambda_i\, m_i' - H \cdot m_i)$$

with $\lambda_i = m_i'^{\mathrm{T}} \cdot H \cdot m_i$, must follow a Ξ^2 square distribution with 2 degrees of freedom. If this hypothesis is rejected with a given probability threshold, we thus can consider that this point correspondence is not related by the given H matrix. This test will be called T_1.

Applications: (1) After the estimation of a H-matrix we can analyse the a posteriori error for each correspondence and reject correspondences which do not correspond to the estimated H-matrix. (2) Similarly, if an initial estimate of the H-matrix is provided we can analyse the a priori error for each correspondence and do not take into account correspondences which are not likely to be related to the expected H-matrix.

T_2: *Is the estimation of a collineation not compatible with the used set of correspondences?* Let us consider a H-matrix estimated from N set of correspondences. The quantity

$$\Xi^2\left(\{m_i, m_i', \Lambda_i^{-1}\}_{i=\{1...N\}}, H\right)$$
$$= \sum_{i=1}^{N}(\lambda_i\, m_i' - H \cdot m_i)^{\mathrm{T}} \cdot \Lambda_i^{-1} \cdot (\lambda_i\, m_i' - H \cdot m_i)$$

must follow a Ξ^2 square distribution with $2\,N - d_H$ degrees of freedom, if the N point correspondences are related by the estimated H matrix, and if the estimated matrix is estimated with d_H degrees of freedom. If this hypothesis is rejected with a given probability threshold, we thus can consider that this set of points is not related by the proposed collineation, but that the motion of these points are governed by another model. This test will be called T_2.

Applications: (1) It allows to decide which representation given in Table 5.1 is to be taken for a H-matrix: we first start the estimate with a minimal numbers of parameters, and if rejected, attempt to use an estimate with more parameters. When different estimates with the same number of parameters are available, the one with the minimal probability of error is chosen.

Please note that we do not have to recompute the normal equations for each case, but only project these equations on the sub-space defined by the constraints.

When the number of points N is high, the quantity $2\,N - d_H \simeq 2\,N$ does not depend on the number of parameters for H, and the estimate with a minimal quadratic error is chosen, which corresponds to a maximal number of parameters. On the contrary, when the number of points N is small, the number of parameters for H influences this criterion. This corresponds to what was required in the previous paragraph.

(2) It also allows us to decide whether the N correspondences are defined by a unique H-matrix or not. If the N correspondences are defined by two or more different H-matrices the statistical test must reject the estimation for any d_H. However, this mechanism is usable only if (a) the two collineations are sufficiently different, with respect to the uncertainties and (b) the two collineations with eventually small parameters cannot be interpreted as a unique collineation with a higher number of parameters.

T_3: *Can two collineations be considered as different?* Let us consider two collineations H_1 and H_2 with their related information matrices I_1 and I_2, respectively. The Mahalanobis distance between these two quantities is given by

$$\Xi^2(H_1, I_{H_1}, H_2, I_{H_2})$$
$$= \text{trace}\left[(H_1 - H_2)^{\text{T}} \cdot (I_{H_1} + I_{H_2}) \cdot (H_1 - H_2)\right] . \qquad (5.37)$$

It must follow a Ξ^2 square distribution with $(d_{H_1} + d_{H_2}) - 1$ degrees of freedom, if the two H-matrices correspond to two estimates of the same collineation. This test will be called T_3.

Application: If this hypothesis is not rejected with a given probability threshold we thus can consider that these two collineations cannot be differentiated, and we thus can optimally merge these two estimations and obtain a fusion of the 2 H-matrices, say H_0, with information matrix is I_{H_0}, using the formulas, to merge H_1 and H_2:

$$\begin{cases} I_{H_0} &= I_{H_1} + I_{H_2} , \\ H_0 &= I_{H_0}^{-1} \cdot (I_{H_1} \cdot H_1 + I_{H_2} \cdot H_2) . \end{cases} \qquad (5.38)$$

Please note that these formulas correspond to over-classical derivations [5.29]. There is however something not obvious here, because – normally – the covariance of a matrix is a tensor. However, the particular form of the covariance obtained in Prop. 5.6 allows to derive these simple formulas which only involve matrices, and we avoid painful tensor algebra.

Analysis of Errors. Let us finally analyse the kind of errors, when some points either do not belong to the right plane or do not have the same rigid motion. For simplicity we assume having the same uncertainty for each point. This limitation has been only chosen to simplify the derivations, whereas similar results could have been obtained in the general case.

Proposition 5.7. If we have at least 3 non-collinear points $\{m_i\}_{i=1...N}$ with their corresponding 3D depth Z_i, and the same uncertainty $\Lambda_i = \sigma^2 I$, a rigid displacement (R_i, t_i) for each point, a camera displacement $(R_\bullet, t_\bullet)$ and a calibration matrix A, the solution to (5.36) is

$$H = \lambda \left[\underbrace{A \cdot R_\bullet \cdot A^{-1}}_{H_\infty} + \underbrace{(A \cdot t_\bullet)}_{s_\bullet} \cdot \mu^{\mathrm{T}} + \epsilon_H \right]$$

with

$$
\begin{cases}
\mu &= \left(\sum_i m_i \cdot m_i^{\mathrm{T}}\right)^{-1} \cdot \left(\sum_i \dfrac{m_i}{Z_i}\right) \\[2mm]
&= \left(\sum_i m_i \cdot m_i^{\mathrm{T}}\right)^{-1} \cdot \left(\sum_i \dfrac{1}{1 + \frac{d(M,P)}{d(O,P)}}\, m_i \cdot m_i^{\mathrm{T}}\right) \cdot \nu \\[2mm]
\epsilon_H &= \left(\sum_i \delta m_i \cdot m_i^{\mathrm{T}}\right) \cdot \left(\sum_i m_i \cdot m_i^{\mathrm{T}}\right)^{-1}
\end{cases}
$$

for any plane P which structure is given by $\nu = (A^{-1\,\mathrm{T}} \cdot n)/d$.

The quantity $\delta m_i = A \cdot (\Delta R_i \cdot M_i + \Delta t_i)/Z_i$ is the retinal disparity (2D displacement) due to the difference between the displacement related to $(R_\bullet, t_\bullet)$ and the effective displacement of the point. It corresponds to the retinal disparity which is due to the addition of an object displacement to the camera displacement.

Proof. For each point M_i undergoing a rigid displacement parameterized by $R = R_\bullet + \Delta R_i$ and $t = t_\bullet + \Delta t_i$, we have

$$M_i' = (R_\bullet + \Delta R_i)\, M + t_\bullet + \Delta t_i \ ,$$

$$\underbrace{\frac{Z_i'}{Z_i}}_{\lambda_i'}\, m_i' = \underbrace{\left(A \cdot R_\bullet \cdot A^{-1}\right)}_{H_\infty} \cdot m_i + \frac{1}{Z_i} \underbrace{A \cdot t_\bullet}_{s_\bullet} + \delta m_i \ .$$

If we introduce this expression in (5.36), we have[5]

$$0 = \sum_i \left(H_\infty m_i + \frac{s_\bullet}{Z_i} + \delta m_i\right) \cdot m_i^{\mathrm{T}} - H \cdot m_i \cdot m_i^{\mathrm{T}} \ ,$$

$$0 = (H_\infty - H) \cdot \sum_i m_i \cdot m_i^{\mathrm{T}} + s_\bullet \cdot \sum_i \frac{m_i^{\mathrm{T}}}{Z_i} + \sum_i \delta m_i \cdot m_i^{\mathrm{T}} \ .$$

Considering the inverse of $\sum_i m_i \cdot m_i^{\mathrm{T}}$, we can write $H = H_\infty + s_\bullet \cdot \mu^{\mathrm{T}} + \epsilon_H$ where μ and ϵ_H correspond to the formula to be demonstrated.

Finally, because for any plane given by n and d and any point M_i

$$
\begin{aligned}
(M_i^{\mathrm{T}} \cdot n) - d &= d(M,P) \\
\Leftrightarrow \quad Z_i\,(m_i^{\mathrm{T}} \cdot \nu) - 1 &= \frac{d(M,P)}{d} \\
\Leftrightarrow \qquad\qquad \frac{1}{Z_i} &= \frac{m_i^{\mathrm{T}} \cdot \nu}{1 + \frac{d(M,P)}{d}}
\end{aligned}
$$

---

[5] To be entirely rigorous, let us mention that we consider the true λ_i as being equal to the estimated $\tilde\lambda_i$. Such an approximation is legitimated by the facts that $\lambda_i \simeq 1 \simeq \tilde\lambda_i$, and $Z_i \simeq Z_i'$.

we have

$$\sum_i \frac{m_i^{\mathrm{T}}}{Z_i} = \left[\sum_i \frac{1}{1 + \frac{d(M,P)}{d(O,P)}} \, m_i \cdot m_i^{\mathrm{T}} \right] \cdot \boldsymbol{\nu}$$

and the two forms proposed for $\boldsymbol{\mu}$ are equivalent.   □

From this proposition we can derive the following properties concerning the errors related to our estimation of H:

- *Error due to an erroneous motion of a point.* This error is proportional, for a given point, to the retinal disparity δ_{m_i} which is due to the addition of an object displacement to the camera displacement. This last quantity is, itself, proportional to the point velocity and inversely proportional to the point depth.
- *Error due to an erroneous location of a point.*
  - *Collineation of the plane at infinity.* The error is inversely proportional to the depth of the point. Moreover, we have experimentally observed, that the quantity $\sum_i m_i/Z_i$ taken over the whole visual field is usually very small, because if the distribution of depth is regular, the bias due to points at a given location of the retina is roughly compensated by the bias due to points at the opposite size of the retina, considering a retinal frame of reference which origin is centered in the image.
  - *Collineation of planes at finite distances.* The error is inversely proportional to the ratio between the Euclidian distance of the point to the plane and the Euclidian distance of the plane to the origin. This result is important because it shows that planar structures whose distances to the origin are small can be very badly estimated. This is coherent with (5.15). On the contrary, planes with a distance to the origin which is very high, in other words, close to the point at infinity will be less sensitive to erroneous locations. This is simply due to the fact that all H-matrices for remote planar structures are very similar and close to H_∞. As a conclusion, *the plane at infinity is not far away.*

5.3.3 Considering Correspondences Between Non-punctual Primitives

This formalism can be extended to the case where correspondences are not established between points but between other tokens.

(1) Let us first discuss a 2D correspondence between a point m on a 2D curve and another point m' on the corresponding 2D curve after a retinal normal displacement related to H. Along a curve, the tangential displacement is generally ambiguous (aperture problem) [5.13], except for points with a high curvature (corners, junctions). Let $\boldsymbol{t}$, $||\boldsymbol{t}|| = 1$, be the tangent of the curve on m, and $\boldsymbol{n}$, $||\boldsymbol{n}|| = 1$, its normal, as illustrated in Fig. 5.4.

In this situation, the planar Euclidian displacement

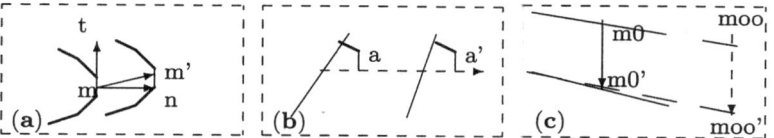

Fig. 5.4. Using correspondences between non punctual primitives; (**a**) correspondence related to the normal motion of a curve, (**b**) correspondence related to the orientation of a line, (**c**) correspondence related to a line segment

$$dm = \frac{m'}{z^T \cdot m'} - \frac{m}{z^T \cdot m}$$

is obtained in the normal direction only, written $\beta = n^T \cdot dm$, whereas $t^T \cdot dm$ is undefined and $z^T \cdot dm = 0$ by definition. Introducing $\lambda m' = H \cdot m$ in the equation involving β we have after a few manipulations:

$$(n^T \cdot m')(m'^T \cdot H \cdot m) = n^T \cdot H \cdot m .$$

The uncertainty for this equation is easily computable. Since only the disparity in the direction of n influences the measure, we can calculate the uncertainty as: $\sigma^2 = n^T \cdot \Lambda \cdot n$, where Λ is the covariance on m'.

Considering a set of normal correspondences we thus can define a criterion, similar to $\mathcal{J}_1'$, to estimate the collineation H which is

$$\mathcal{J}_3' = \sum_i \frac{1}{\sigma_i^2} \left\| (n_i^T \cdot m_i')(m_i'^T \cdot H \cdot m_i) - (n_i^T \cdot H \cdot m_i) \right\|^2 .$$

However, if we define

$$\Lambda_i'^{-1} = \frac{n_i \cdot n_i^T}{\sigma_i^2} ,$$

i.e. if we consider that the variance is infinite in the tangential direction, this criterion is equivalent to the criterion $\mathcal{J}_1'$, used in Prop. 5.5. Therefore *we can introduce 2D correspondences related to the normal displacement of a 2D curve, in our framework, without any change, but using an appropriate information matrix.*

(2) Similarly, a 2D line direction of angle a in correspondence with a 2D line direction of angle a' (represented by a pair of vanishing points) yields the same equation as for a point. If we use a weaker constraint and simply require the orientation to be preserved, we only have to constraint a vanishing point to map on a finite or infinite point with the required orientation. This corresponds to

$$\begin{vmatrix} \lambda \cos(a') \\ \lambda \sin(a') \\ \mu \end{vmatrix} = H \cdot \begin{vmatrix} \cos(a) \\ \sin(a) \\ 0 \end{vmatrix} \tag{5.39}$$

for some λ, μ (see diagram below). Note that the mapping can not be a rotation of the plane if $\mu \neq 0$. Such a constraint is related to the retinal displacement of line-segments as used in [5.34].

Again, we can introduce such a measurement by simply choosing an information matrix of the form $\Lambda_i'^{-1} = \left(n_i \cdot n_i^{\mathrm{T}} \right) / \left(\sigma_i^2 \right)$ with $n_i = (-\sin(a')$, $\cos(a'), 0)^{\mathrm{T}}$ which effect is to eliminate λ and μ in the previous equation.

(3) A 2D line D with equation $m \in D \Leftrightarrow 0 = n^{\mathrm{T}} \cdot m$ represented by an homogeneous vector n in correspondence with a 2D line D' represented by an homogeneous vector n' yields $\lambda n = H^{\mathrm{T}} \cdot n'$, for some λ. Up to a duality, it yields the same linear equation as for point correspondences. Moreover we can reduce this equation to equations involving points, as developed now.

Let us choose two points m_0 and m_1 on this 2D line. In practice, we are going to detect line-segments in an image. It has been shown [5.17] that we better choose the center of the segment m_0 and the direction of the segment, i.e. the vanishing point, or point at infinity $m_1 = m_\infty$. This will minimize the covariance of the measures. However, any other pair of points could be used.

The point m_0 is in correspondence with a point $m_0' + \lambda_0 l'$, where l' is the direction of the line and λ_0 some constant, since we cannot determine the exact location of this correspondence in D' but only its location up to a displacement on the line. In other words, we match m_0 with a "sliding point". Similarly m_1 is in correspondence with a point $m_1' + \lambda_1 l'$ for some λ_1. We then have $m_0' + \lambda_0 l' = H \cdot m_0$ and $m_1' + \lambda_1 l' = H \cdot m_1$. But we have $n' = k' \, m_0' \wedge m_1'$ since this pair of points generates the line D'. Moreover, we can derive[6]

$$
\begin{aligned}
n &= k \, m_0 \wedge m_1 \\
&= \frac{k}{\det(H)} \, H^{\mathrm{T}} \cdot (H \cdot m_0 \wedge H \cdot m_1) \\
&= \frac{k}{\det(H)} \, H^{\mathrm{T}} \cdot [(m_0' + \lambda_0 l') \wedge (m_2' + \lambda_2 l')] \\
&= \frac{k}{\det(H)} \, H^{\mathrm{T}} \cdot \left(\frac{1}{k'} n' + \lambda_0 m_0' \wedge l' - \lambda_2 m_2' \wedge l' \right) = \rho \, H^{\mathrm{T}} \cdot n'
\end{aligned}
$$

since, using the fact that $l' = \kappa \, [m_0' - m_2']$ we easily verify the last equality while we use

$$
\rho = \frac{k}{\det(H)} \left[\frac{1}{k'} + \kappa(\lambda_2 - \lambda_0) \right] .
$$

In other words, if we have a line correspondence, *constraining the collineation using the equation on the line parameter n or using the normal*

---

[6] We make use of the following relation:
$$
(M \cdot x) \wedge (M \cdot y) = \underbrace{\det(M) \, M^{-1\,\mathrm{T}}}_{M^*} \cdot (x \wedge y) .
$$

correspondences between two points on this line is equivalent. We thus do not
need to develop specifically the case of lines, but can limit our developments
to the case of points, with appropriate information matrices.

More generally, correspondences between points of interests, correspondences related to a line-segment and correspondences between points along a
curve can be combined into a single equation since the contribution of each
point is additive in (5.36).

In fact, we can combine a lot of different sets of tokens as illustrated in
the next table:

Object(s) in correspondence	Number of equations for H
Normal motion	1
One point	2
One vanishing point	2
One line	2
One line segment	3
Two distinct points	4
A corner	4
One line, one point	4
Three non-collinear points	6
etc.	...

Hence the H-matrix can be estimated considering points, line-segments,
local orientation or normal displacements as primitives. It is however not
straightforward to integrate other algebraic primitives. For instance, a 2D
conic C with equation $m \in C \Leftrightarrow 0 = m^{\mathrm{T}} \cdot C \cdot m$ thus represented by an
homogeneous symmetric matrix C in correspondence with a 2D conic C'
represented by an homogeneous symmetric matrix C' yields $\lambda\, C = H^{\mathrm{T}} \cdot C' \cdot H$,
for some λ. In this case the equations are quadratic, and we cannot reduce
this equation to some set of linear equations. We thus will not attempt to
use such objects here.

5.4 Implementation and Experimental Results

Let us now apply the previous piece of theory to some effective modules
of computation of retinal motion in the uncalibrated case. The goal of this
section is by no means to extensively experiment *all* that has been developed
in the previous sections, but to verify some of the main points in order to
check the validity of our approach.

The Early-Vision Module. In order to obtain point correspondences, we have detected corners in one image considering a technic similar to [5.9] and have established correspondences using a simple correlation operator: this correlation operator is used under the hypothesis that the disparity between two .nsecutive images is small and works as follows.

The chosen criterion is the sum of the absolute values of the intensity errors taken in a small window (typically 7×7) centered at the point locations in both images. For a given point location in one frame, we start at the same location in the previous frame and, using a gradient descent, try to find the first local minimum of the criterion. The shape of the correlation is used to estimate the covariance of estimate, and some tests are performed to validate the correspondences. All these implementation details have been already described for a similar operator used in stereo [5.35], and we have re-used the same technic to get subpixel precision, reject false matchings and estimate the uncertainty. The obtained precision is a bit better than one pixel.

Estimation of a Collineation from Point or Token Correspondences. The proposed algorithm is a straightforward application of the previous section, and has been implemented as follows:

– *Input:*
  – Information matrix $\{m_i, m_i', \Lambda_i^{-1}\}_{i=1...N}$ and a set of point correspondences.
  – An initial value of the collineation and the related information matrix (H, I_H), with $H = 0$ and $I_H = 0$ if no prior information is given on H.
  – A probability threshold to reject outliers.
– *Algorithm:*
  (1) Reject all points which are not compatible with the initial value of the H-matrix, if any, using the test T_1.
  (2) Compute the coefficients of the function $f(...)$ given in (5.36) of Prop. 5.5.
  (3) For each model, described in Table 5.1:
    – Estimate the H using (5.36).
    – Using the Test T_2 decide if the model is acceptable.
    – Choose the best model, according to T_2.
  (4) Repeat from Step 1, until the set of points not compatible with the value of H-matrix does not vary, and while the result of the test T_2 improves at each iteration.
– *Output:*
  – A new estimation of H and its information matrix.
  – A set of points which retinal motion is compatible with H.

This algorithm can integrate not only point correspondences, but also correspondences between other primitives as discussed previously. It not only calculates the coefficients of the collineation but also detect if the collineation is likely of a particular kind. It is capable of rejecting some outliers, so has

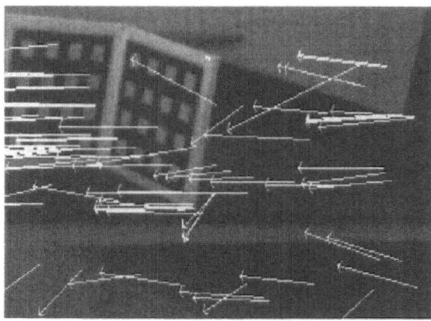

Fig. 5.5. Using point correspondences to estimate a collineation. The estimated retinal motion is shown in the *left image* and the residual error is shown in the *right image*. It is clear that the collineation of the *left plane* has been estimated, despite the presence of a lot of outliers. The displacement vectors have been drawn with a magnitude factor of 10

some properties of robustness [5.36] and the convergence is achieved if the number of outliers and their bias are small enough, as discussed in [5.29].

Such an example is shown in Fig. 5.5, in which we started the estimation from scratch, for an image containing several planes. The residual error was 1.053 pixel in average (standard deviation).

Since the precision of the estimate is such that the residual error, for each point of the right plane is less than 1 pixel in average, which is close to the precision of the early-vision module, the estimation can be considered as almost optimal.

Although the estimation could have "oscillated" between several planes, it did not and the convergence is obtained for the plane with a maximal number of matches. By eliminating the matched points we can very easily estimate the collineation of another plane. This is shown in Fig. 5.6.

We can formalize such a method as follows.

Segmentation of the Motion Field Using Collineations.

- *Input:*
    - Information matrix $\{m_i, m'_i, \Lambda_i^{-1}\}_{i=1...N}$ and a set of point correspondences.
    - A probability threshold to reject outliers.
- *Algorithm:*
(1) Estimate a first H-matrix taking all points into account.
(2) Memorize points which retinal motion is not compatible with this estimate.
(3) Repeat step 1, estimating a new H-matrix, using the previous set of points, until all points except at most 4 have been attached to a H-matrix.

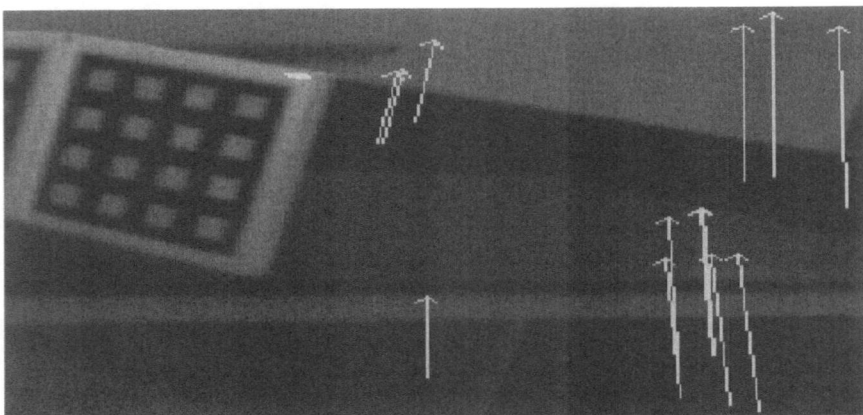

Fig. 5.6. Detecting several collineations in an image. The first identification has been shown in the *previous figure*, and the present identification corresponds to points in the *dark almost vertical* plane. The estimated displacement vectors are drawn with a magnitude factor of 10

– *Output:*
  – A set of H-matrix with their information matrix and a list of points compatible with each collineation.

   This algorithm of segmentation is very simple but still efficient in a complex scene such as a scene in which an object is in motion, as shown in Fig. 5.7, or for which we have several planar structures , as shown in Fig. 5.6.

Parametrization of the Retinal Motion in an Image Sequence. Let us now briefly discuss the problem of a sequence of images in which several retinal motions are present. We are going to make the following two assumptions about the visual system and the environment:

H1: The intrinsic parameters of the camera are locally constant, i.e. have very slow variations.

H2: The self-motion and the object motions are locally performed at almost constant velocity, i.e. with negligible accelerations.

As a consequence *we can assume that the H-matrices are locally constant.* We thus can use the previous estimation of a H-matrix as the initial estimate for the next frame.

   The parametrization of the retinal motion is initialized (bootstrapping) using the previous module and can be updated as follows:

– *Input:*
  – A set of collineations and their information matrix estimated in the previous frame.

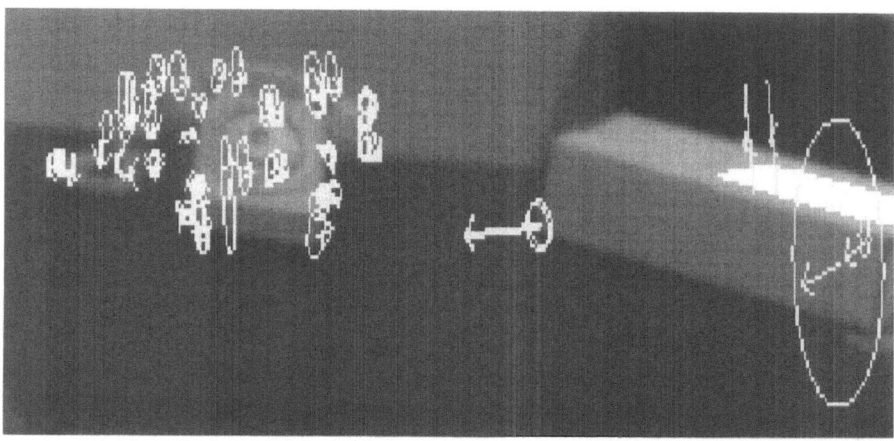

Fig. 5.7. Detecting collineations in the presence of a moving object. The *white box* on the *right* is not stationary but translated on the table. The *lower left corner* of the box, for instance, is clearly detected as a non-stationary token. The residual displacement vectors with a magnitude factor of 10 and the ellipsoid of uncertainty with a magnitude factor of 100 are shown

- Information matrix $\{m_i, m_i', \Lambda_i^{-1}\}_{i=1\ldots N}$ and a set of point correspondences, plus the index of the collineation compatible with the previous retinal correspondence.
- A probability threshold to reject outliers.
- *Algorithm:*
(1) For each H-matrix update its estimation using the points with motion is compatible with the previous estimate.
(2) For each point, using the test $\mathcal{T}_1$ find the H-matrix which compatibility with the retinal motion is optimum.
(3) For points not associated with a H-matrix (new matches, points close to the image borders, etc.) apply the last procedure to segment these points in sets of compatible retinal motions.
(4) If two H-matrices are compatible, as stated by the test of (5.37), perform the fusion of the two estimates as in (5.38).
- *Output:*
  - A set of H-matrix with their information matrix and list of points compatible with the estimated collineation.

This mechanism allows very easily to integrate several frames and to improve the representation of the retinal motion with time.

Recovering Euclidian Parameters of the Scene. Considering that H_∞ has been identified (see Sect. 5.2.2), we can very easily recover using equations developed in Sect. 5.2, either global parameters such as:

- the matrix H_∞ and its information matrix.

- the focus of expansion $s_\bullet$ and its information matrix,
- the five parameters of the matrix K defined in (5.30) and their related information matrix,
- the five intrinsic parameters of the matrix A and their related information matrix;

and for each planar structure:

- its rigid motion,
- the "uncalibrated" structure ν and its information matrix,
- the Euclidian parameters n/d and its information matrix.

These developments are out of the scope of this book and represent future work, but we however have manually computed a few of them, and have compared, when available, with known values.

Let us first consider, for instance, focus of expansion. In image sequences in which the displacement is not only a translation, we would like to compute the focus of expansion. We have compared our method with a well established method developed by [5.20] and robustly implemented by [5.37]. This has been considered as the reference value. The estimated motion field is shown in Fig. 5.8, for both image sequences.

We have obtained the following results:

	Obtained Value	Reference value
First Sequence	(205, 240)	(210, 232)
Second Sequence	(181, 230)	(160, 200)

which are coherent, but the lack of other reference methods to compute such parameters in the uncalibrated case, does not allow us to develop these experimentations in more detail for a general scene.

However, in order to control the experimental parameters we have experimented our module on a robotic head as described in [5.25, 38]. We have observed a simple scene with two planes and have analysed the retinal displacement attached to each plane. Results are shown in Fig. 5.9.

Since on such a system we know the calibration parameters

$$u_0 = 247.4 , \quad v_0 = 267.6 , \quad a_u = 523.3 , \quad a_v := 747.8$$

and the characteristics of the displacement we can predict the collineation in function of $n/d = (n1, n2, n3)^{\mathrm{T}}$. We have obtained:

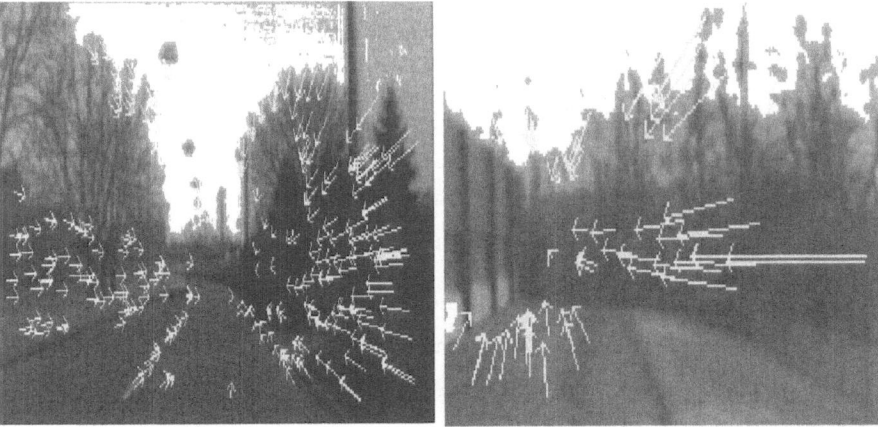

Fig. 5.8. Two examples of motion computations used to calculate focus of expansions

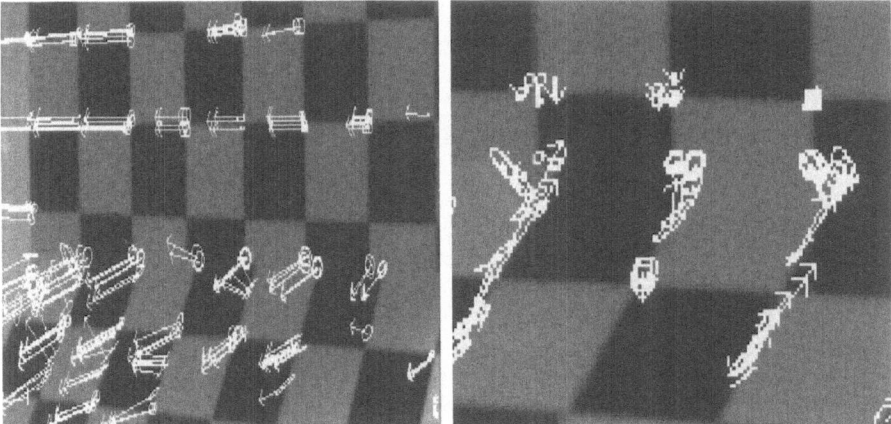

Fig. 5.9. Detecting collineations to recover the Euclidian structure of the scene. The estimated retinal motion is shown in the *left image*, with covariances represented as ellipsoids or error. Magnitudes are the same as previous figures. The residual error for the upper plane is shown in the *right image*

$$
H(n1,n2,n3) = \begin{pmatrix} 0.984+0.008\,n1 & -0.121+0.016\,n2 & 3.758-2.140\,n1+4.527\,n3 \\ 0.2481+0.001\,n1 & 0.984+0.001\,n2 & -61.394-0.391\,n1+0.8277\,n3 \\ 0 & 0 & 1.0 \end{pmatrix}.
$$

This numerical example shows that we can easily estimate $n1$ and $n3$ and in a less important proportion $n2$ because their contribution to the coefficients of H are not negligible in this experimental set-up.

We have measured

$$\hat{H} = \begin{pmatrix} 0.96 & -0.12 & 25 \\ -0.0067 & 0.95 & 5.2 \\ 0.00003 & -0.0002 & 1 \end{pmatrix}$$

and have obtained

```
n:= [ -.917, -.0889, .386 ] with d:= 0.1149
```

which is close to the expected values $n = (-0.91, 0, 0.4)$ and $d = 0.1$ for the vertical plane.

Similarly we have obtained for the other plane

```
n:= [ -.365, -.912, .182 ] with d:= 0.8234
```

for expected values of $n = (-0.3, -0.9, 0.2)$ and $d = 0.8$ which is quite accurate.

Finally we have analysed three planar structures of an outdoor scene shown in Fig. 5.10. We have considered the road as being a kind of "U" as shown in Fig. 5.10 also. We thus have to detect three main planar structures the road, left border and right border.

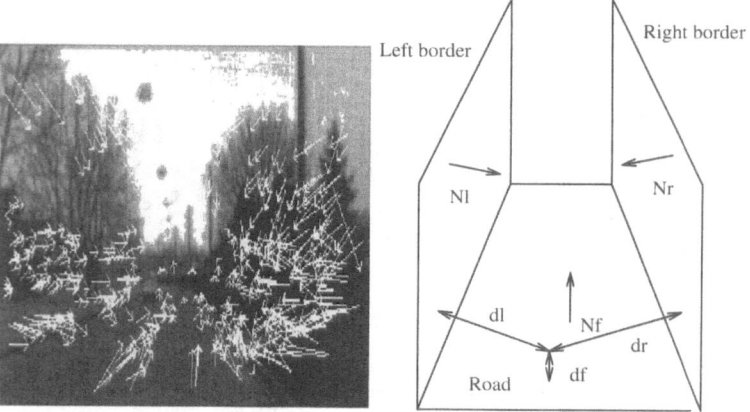

Fig. 5.10. Using an outdoor scene to manually extract the main planar structures. The image is shown on the *left* and the simple model on the *right*

This has been done considering windows in the image which do not need to be accurately estimated because the estimation is robust and automatically eliminates outliers. The result is shown in Fig. 5.11 and one example of output, for the plane corresponding to the road, follows

```
xi2=5.60034 xi4=2.73751 xi6=2.81213 xi8=1.13952
        (       0.99      -0.063        0.31 )
H =     (    -0.00077       0.99        0.24 )
        (    -2.5e-05    -0.00027          1 )
```

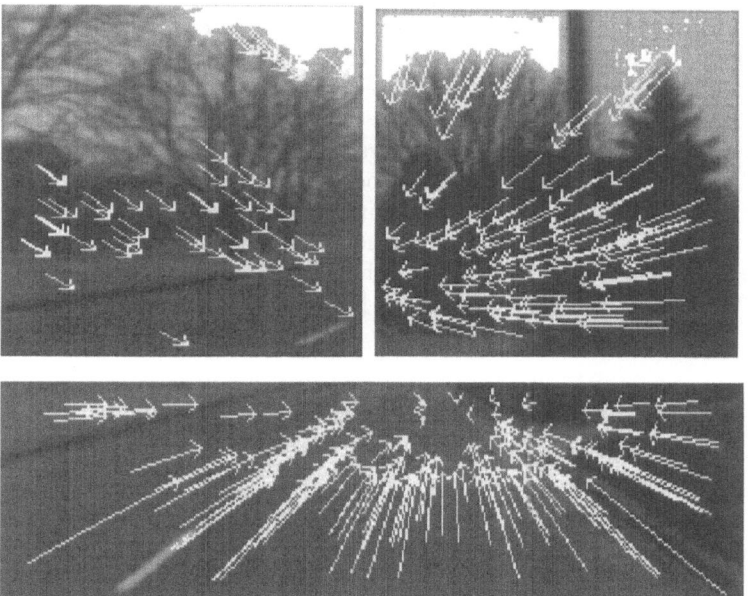

Fig. 5.11. The motion field estimated for the three main planar structures

Please note that xi2, xi4, xi6, and xi8 correspond to the weighted mean-square errors for some models shown in Table 5.1 (constant field, 1D affine, affine and homographic), the best being the last one in this case.

We then have evaluated the order of magnitude of the distances, using approximate values for the calibration and self-motion, and have obtained

dl = 12 meters, dr = 4 meters, df = 2 meters

which are not unrealistic values. We would like to stress that this last experiment is by no means a quantitative evaluation of the performances of the algorithm, but a qualitative test that "it still works for outdoors scenes".

5.5 Conclusion

Our goal was to analyse an image sequence using a set of collineations with the goal of recovering the piece-wise planar structures of the scene. This approach is often followed in computer vision but we have taken into consideration several new aspects of this problem: (1) An uncalibrated image sequence, (2) The integration of several images in the same representation, (3) The detection of moving objects.

We have in a certain sense avoided the problem of image segmentation which is often address in this case, because we have a statistical model which tolerates that some points can be temporary attached to a wrong patch.

We have introduced some robustness in the estimation of the retinal displacements, with the capability to switch between different representations of the local retinal motion, depending on the data.

Finally, the proposed equations show that there is a compromise between pure projective – or uncalibrated – parameters and Euclidian reconstruction. It is definitely possible to analyse many aspects of the image sequence by only using a projective representation, and our set of collineations is – maybe – the simplest one. But it allows us to incrementally recover some projective and/or affine parameters of the scene, and then, Euclidian data. Therefore, Geometry is not only amazing and beautiful [5.13], it is also useful.

6. Uncalibrated Motion of Points and Lines

As required in any active vision system, we address the problem of computing structure and motion, given a set of points and/or line correspondences, in a monocular image sequence, when the camera is **not** calibrated.

Most of the earlier or recent studies in the field assume that the calibration of the system is known (far from exhaustive bibliography being [6.1–4]), except for [6.5–8]. However, these scientists have studied only the case of points correspondences, and have restricted their approach to the case where the intrinsic parameters of the camera are constant, while only 2 or 3 views have been taken into account. The generalization of the case where points and lines are given has not been made (except [6.9, 10] for some results), and the problem of considering non-constant intrinsic parameters has also not yet been addressed.

This is however an important challenge. In particular, in the case of active vision, the extrinsic parameters and the intrinsic parameters of the visual sensor are modified dynamically. For instance, when tuning the zoom and focus of a lens, these parameters are modified and must be considered as dynamic parameters. It is thus relevant to attempt to determine dynamic calibration parameters by a simple observation of an unknown stationary scene, when performing a rigid motion.

6.1 Introduction

The facility in a motion paradigm is that we can assume the disparity between two frames to be small, leading to easy solutions for the correspondence problem. These correspondences can efficiently be established (token-tracking) [6.11, 12]. The problem is thus, given correspondences between points or lines, to recover, the motion, structure and calibration of the system.

Let us first emphasize the generality of the approach we want to take here: in most of the artificial vision problems or "paradigms", four quantities are to be manipulated (Fig. 6.1): (1) the projection from the 3D world onto the camera, (2) the correspondences between tokens in two cameras (either at different locations in space such as for stereo, or at different times such as for motion analysis), (3) the rigid displacement or motion (either in space or time) and (4) the 3D structure.

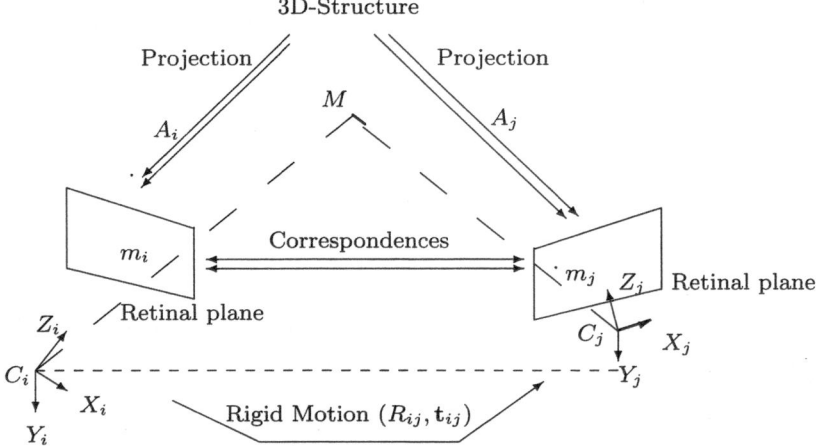

Fig. 6.1. Elements used in the definition of points in motion

Let us now summarize, for different vision paradigms, what is given to the algorithm (input)[1] and what is estimated by the algorithm (output):

Paradigm	Projection	Correspondences	Motion	3D Structure
Structure from Motion	input	input	input	output
Stereo Matching	input	output	input	not used
Stereo Reconstruction	input	input	input	output
Pose determination	input	input	output	input
Motion Computation	input	input	output	not used
Token Tracking	not used	output	output	not used
Off-line Calibration	output	input[1]	input	input

In every algorithms, only one quantity is output, whereas all others are input or not used. However, we do not think this is the right way, because these quantities are in deep interaction. We thus would like to treat the problem in the following manner:

---

[1] In the case of "off-line calibration" a correspondence means a correspondence between the 2D point in one image and the 3D model used for calibration.

Paradigm	Projection	Correspondences	Rigid Displacement	3D Structure
Motion Without Calibration	output	input	output	output

This corresponds also to what is called "auto-calibration" [6.5, 8, 13] since we have introduced the calibration parameters as unknowns in the state of the system.

In order to attain this objective, this chapter is divided into three parts:

In the first section, we attack the problem of point correspondences only, and derive the equations related to the motion of points, depending on the chosen geometry, for a system of cameras without calibration.

In the second section, we attack the problem from a geometric point of view, considering now point and line correspondences and generalize the epipolar geometry used for points to a new construction, called trifocal geometry, which allows us to solve the motion problem in three views.

In the final section, we propose an implementation and report some experimental results.

Notations. We note vectors using bold letters and matrix using capital letters. The duals of vectors are represented as the transpose of a vector and scalars are in italic. The notation $x \wedge y = \tilde{x} \cdot y$ corresponds to the cross-product, the dot-product being written as $x^T \cdot y$. The identity matrix is written I. Geometric objects such as points, lines, planes are written with capital letters in 3D, and small letters in 2D.

6.2 Representations of the Retinal Motion for Points

6.2.1 Considering the Euclidean Parameters of the Scene

Camera Model and Frame of Reference. We use *the standard pinhole model* for a camera, in a position indexed by i, assuming the camera performs a perfect perspective transform with center C_i (the camera optic center) at a distance f (the focal length) of the retinal plane.

It must be noted, that the pinhole model can still be used for a zoom lens if the object-to-image distance is not considered as fixed [6.14]. This is the case here, since we will adapt the camera metric for different object locations.

All coordinates are related to an affine frame of reference $\mathcal{R}_i = (C_i, \, x_i, \, y_i, \, z_i)$ *attached to the retina*, z_i being aligned with the optical axis, x_i and

y_i being aligned with horizontal and vertical axis in the image. The retinal plane is thus perpendicular to the optical axis $C_i z_i$. All quantities attached to frame i are indexed by i also.

We represent a 3D point $M_i = C_i M = (X_i, Y_i, Z_i)^{\mathrm{T}}$ using Euclidean coordinates. Points onto the retina, with horizontal and vertical coordinates (u_i, v_i) in pixel will be represented as homogeneous 3D vectors: $s\, m_i = s\, C_i m = (s\, u_i, s\, v_i, s)^{\mathrm{T}}$, corresponding to lines of a given direction passing through the optical center (2D projective space). Algebraic developments will thus involve 3D vectors in both cases.

In this study, *we must not assume the system is calibrated*. However, we are in a specific situation because we have chosen a "canonical" frame attached to the retina. Therefore, we consider only the matrix of the intrinsic parameters (called A-matrix) in the projection. Moreover, we have chosen a representation in which points at finite distances are represented their usual Euclidean coordinates, while points in the plane at infinity are represented in homogeneous coordinates. Considering the usual notations with which a 3D point in homogeneous coordinates projects onto a retinal point of 3 homogeneous coordinates using a 4×3 matrix, we can write:

$$
\begin{aligned}
Z_i\, m_i &= \begin{vmatrix} Z_i\, u_i \\ Z_i\, v_i \\ Z_i \end{vmatrix} \\[6pt]
&= \left(\underbrace{\begin{pmatrix} \alpha_u & \gamma & u_0 \\ 0 & \alpha_v & v_0 \\ 0 & 0 & 1 \end{pmatrix}}_{A_i} \quad \begin{matrix} 0 \\ 0 \\ 0 \end{matrix} \right) \cdot \begin{vmatrix} X_i \\ Y_i \\ Z_i \\ \mathcal{U} \in \{0,1\} \end{vmatrix} \\[6pt]
&= A_i \cdot M_i\, .
\end{aligned}
\tag{6.1}
$$

Please note that we assume that the intrinsic parameters are different for each camera position, although we have not indexed the A-matrix components for simplicity.

This corresponds, if $\gamma = 0$, to the usual equations $u = \alpha_u X/Z + u_0$, $v = \alpha_v Y/Z + v_0$. See [6.15] for a recent review. In particular $\det(A_i) = \alpha_u \alpha_v = k/f^2$ is proportional to the inverse of the square of the focal length.

For points not at infinity, we have $\mathcal{U} = 1$ and Z_i corresponds to the Euclidean depth with this model. Equation (6.1) is still valid for "points at infinity", i.e. vanishing points, with $\mathcal{U} = 0$, but Z_i does not correspond anymore to the depth of the point. Therefore, our notation is suitable to represent both Euclidean and projective quantities of the 3D space. Moreover, and contrary to the usual notation involving a 3×4 matrix, the introduction of the A-matrix will simplify the algebraic derivations in the sequel.

Representation of Rigid Displacements. We consider only the motion of a unique rigid object or the self-motion of the camera observing a stationary scene, *in the discrete case*. We thus represent motion through rigid displacements.

It means that the tokens in the scene are undergoing a rigid displacement parameterized by a rotation matrix R_{ij} and a translation vector t_{ij}:

$$M_i = R_{ij} \cdot M_j + t_{ij} \ . \tag{6.2}$$

This equation defines the 3D correspondence, for a point, between two frames. The reverse relation is $M_j = R_{ij}^{\mathrm{T}} \cdot M_i + t_{ji}$ and we obviously have $R_{ji} = R_{ij}^{\mathrm{T}}$ and $t_{ji} = -R_{ij} \cdot t_{ij}$. Moreover rotations and translations form the well-known group of rigid displacements for which we have $R_{ij} = R_{ik} \cdot R_{kj}$ and $t_{ij} = t_{ik} + R_{ik} \cdot t_{kj}$, while $R_{ii} = I$ and $t_{ii} = 0$.

Therefore, as soon as we know the translations and the rotations between consecutive pairs of views, we can calculate rotations and translations between any other pairs of views.

The Euclidean Representation of Motion in an Image Sequence. As a consequence of the previous discussions, when the system is not calibrated, considering $N + 1$ views, we have:
(a) N set of parameters related to the rigid motion between two consecutive views and
(b) $N + 1$ set of parameters related to the calibration, while
(c) the motion is known up to a scale factor, since we consider a monocular image.

So, the retinal displacement is a function of:

$$3\,N\,(\text{3D rotation}) + 3\,N\,(\text{3D translation})$$
$$+5\,(N + 1)\,(\text{intrinsic parameters}) - 1\,(\text{scale factor})$$
$$= \quad 11\,N + 4\,(\text{parameters}) \ .$$

This representation is schematized in Fig. 6.2.

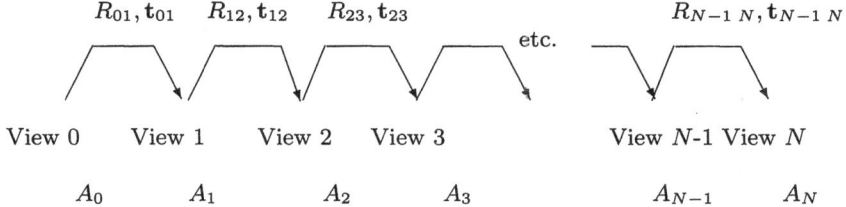

Fig. 6.2. Using calibration parameters (*A*-matrices) and rigid displacements (*R*-matrices and *t*-vectors) to represent the retinal motion in an image sequence

With this parametrization, considering the point location in the first view and its depth, we can recover its location the N other views. Moreover, we can easily calculate its depth considering the point location in the first two views (structure from motion), as it is going to be made explicit for the next representation, and is well known in the Euclidean case [6.16]. As a consequence, with this parametrization, considering the point location in the first two views, we can recover its location the $N - 1$ other views.

It corresponds to a Euclidean parameterization of the retinal motion. However, although we have a precise definition for calibration and motion, these parameters are not directly related to what is measured in an image sequence: the 2D correspondences and to what is usually to be recovered: the depth of the 3D points. This is going to be done using a special representation, the Qs-representation in the next subsection.

6.2.2 From Euclidean to Affine Parameters

Definition and Properties of the Qs-Representation. Considering the 2D correspondences between two points m_i and m_j in two different frames, we obtain, combining (6.1) and (6.2):

$$Z_i \, m_i = Z_j \underbrace{A_i \cdot R_{ij} \cdot A_j^{-1}}_{Q_{ij}} \cdot m_j + \underbrace{A_i \cdot t_{ij}}_{s_{ij}} . \tag{6.3}$$

The quantity Q_{ij} corresponds to the "uncalibrated rotational component of the rigid displacement". In fact, this matrix corresponds to *the collineation of the plane at infinity from frame j to frame i*, as easily established considering points with $Z_i \to +\infty$, or remembering that for points at infinity (in practice at the horizon) the translational component of the motion is not to be taken into account (in practice is negligible). If the calibration parameters are constant, i.e. $A_i = A_j = A$ then $\det(Q_{ij}) = 1$ but in any case

$$\det(Q_{ij}) = \frac{\det(A_i)}{\det(A_j)} \simeq \left(\frac{f_j}{f_i}\right)^2$$

is directly related to the variation of the focal length of the system, i.e. the "zoom". However, despite the fact this collineation is defined up to a scale factor, the algebraic quantity Q is entirely defined, with our notations.

This matrix is in fact related to the affine geometry of the scene, because the displacements of 3D line directions, i.e. vanishing points, are entirely defined by Q_{ij}. For instance the reader can easily verify that two 3D lines are parallel if and only if their intersection in the frame j is mapped onto their intersection in the frame i by Q_{ij}.

The quantity s_{ij} corresponds to the "uncalibrated translational component of the rigid displacement", also called "focus of expansion" by some scientists. In fact, this vector corresponds to the epipole in frame i, i.e. *the projection of the optical center of frame j onto the retinal plane R_i of frame i*, since it is the image of the quantity $m_j = (0, 0, 0)^{\mathrm{T}}$. However, despite the fact it is in relation with an homogeneous quantity, the magnitude of this vector is entirely specified, with our notations.

Considering the previous definitions, and the relation between rotations and translations, we have the following obvious relations between collineations and epipoles:

$$
\begin{array}{llll}
Q_{ii} & = & I & ; \quad s_{ii} = 0 ; \\
Q_{ij} & = & Q_{ji}^{-1} & ; \quad s_{ij} = -Q_{ij} \cdot s_{ji} ; \\
Q_{ij} & = & Q_{ik} \cdot Q_{kj} & ; \quad s_{ij} = s_{ik} + Q_{ik} \cdot s_{kj} .
\end{array}
\tag{6.4}
$$

Then, in an image sequence, as soon as the Qs-representation between two consecutive views is known, any other Qs-representation can be estimated.

This representation is defined up to two scale-factors as it is easily seen, when making explicit (6.2):[2]

$$
\left\{
\begin{cases}
u_i = \dfrac{Q_{ij}^{00} u_j + Q_{ij}^{01} v_j + Q_{ij}^{02} + \dfrac{1}{Z_j} s_{ij}^0}{Q_{ij}^{20} u_j + Q_{ij}^{21} v_j + Q_{ij}^{22} + \dfrac{1}{Z_j} s_{ij}^2} , \\[3em]
v_i = \dfrac{Q_{ij}^{10} u_j + Q_{ij}^{11} v_j + Q_{ij}^{12} + \dfrac{1}{Z_j} s_{ij}^1}{Q_{ij}^{20} u_j + Q_{ij}^{21} v_j + Q_{ij}^{22} + \dfrac{1}{Z_j} s_{ij}^2} ,
\end{cases} \\
\dfrac{1}{Z_i} = \dfrac{1}{Z_j} \dfrac{1}{Q_{ij}^{20} u_j + Q_{ij}^{21} v_j + Q_{ij}^{22} + \dfrac{1}{Z_j} s_{ij}^2} .
\right.
\tag{6.5}
$$

If we multiply every depth Z_i and both Q_{ij} and s_{ij} by a constant factor λ these equations are left unchanged. This "expansion" factor controls the relative scale between frame i and frame j. This indetermination disappears if the determinant of Q_{ij} is known, that is if the variation of the focal length if known, because this expansion corresponds to the "relative zoom" between two frames. It is – a priori – not known if the calibration is not known.

In addition to that, if we multiply every depth Z_i, Z_j and s_{ij} by another constant factor μ in these equations, they are also left unchanged. This "scale" factor corresponds to the usual scale factor indetermination observed in a monocular image sequence.

There are thus two kinds of scale factors:

– **Local** scale factors control relations between two consecutive frames and vary from one pair of frames to another. The "expansion" factor identified previously is one example.
– **Global** scale factors are unique for the whole sequence and correspond to a global attribute. The "scale" factor identified previously is one example.

Reciprocally, because these equations are linear with respect to Q and s, these are the only transformations which leave these equations unchanged,

---

[2] We represent the components of a matrix or a vector using upper subscripts from 0 to 2.

considering a enumerable set of generic points. So, $9\,(Q\text{-components})\,+$
$3\,(s\text{-components})\,-\,2\,(\text{scale-factors})\,=\,10\,(\text{parameters})$ if considering only
two views, but when considering an image sequence, relating the depths be-
tween two consecutive views, only one scale factor remains, the other being
global for the whole sequence.

In order to avoid this indetermination we can choose different constraints,
depending on the nature of the problem. For instance, if the intrinsic calibration parameters are constant, i.e. if $A_i = A_j$, we have $\det(Q_{ij}) = 1$. Moreover,
if we choose $\|s_{ij}\| = 1$ both indeterminations are avoided.

The Affine Representation of Motion in an Image Sequence. As a consequence of the previous discussion, when the system is not calibrated, considering $N + 1$ views and the Qs-representation, we have: (a) N set of 12
parameters related to the 9 components of each Q-matrix and the 3 components of each s-vector, while (b) these parameters are known up to N local
scale factors, and one global scale factor.

Then, in a sequence of $N + 1$ views, the Qs-representation is defined,
considering correspondences in the whole sequence, the 2D correspondences,
by $11\,N - 1$ parameters.

This representation is schematized in Fig. 6.3.

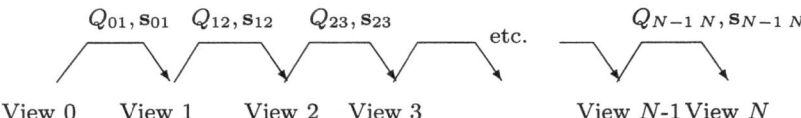

Q_{01}, s_{01} Q_{12}, s_{12} Q_{23}, s_{23} etc. $Q_{N-1\,N}, s_{N-1\,N}$

View 0   View 1   View 2   View 3   View N-1 View N

Fig. 6.3. Using the Qs-representation to represent the retinal motion in an image
sequence

With this parametrization, considering the point location in the first view
and its depth, we can recover its location in the N other views. Moreover, we
can easily calculate its "affine depth" considering the point location in the
first two views, as made explicit soon, in Prop. 1. As a consequence, with this
parametrization, considering the point location in the first two views, we can
recover its location the $N - 1$ other views.

It corresponds to an affine parameterization of the retinal motion.

Comparing the previous representation, 5 parameters are missing. This
comes from the fact that we have not made the calibration parameters explicit, but have mixed them with the motion parameters. Let us analyse this
situation.

Relation Between the Qs-Representation and Euclidean Parameters. In fact,
the Q-matrix is in deep relation with intrinsic parameters, the A-matrix, and
the rotational component of the displacement. Let us explicit this relation.

If we write that $R_{ij} = A_i^{-1} \cdot Q_{ij} \cdot A_j$ is an orthogonal matrix, i.e. $R \cdot R^{\mathrm{T}} = I$, we obtain, after some algebra:

$$Q_{ij} \cdot K_j \cdot Q_{ij}^{\mathrm{T}} = K_i \quad \Leftrightarrow \quad Q_{ij}^{T} \cdot K_i^{-1} \cdot Q_{ij} = K_j^{-1} \tag{6.6}$$

where

$$K_i = A_i \cdot A_i^{\mathrm{T}} = \begin{pmatrix} \underbrace{\alpha_u^2 + \gamma^2 + u_0^2}_{b_u} & \underbrace{u_0\,v_0 + \gamma\,\alpha_v}_{b_c} & u_0 \\ \underbrace{u_0\,v_0 + \gamma\,\alpha_v}_{b_c} & \underbrace{\alpha_v^2 + v_0^2}_{b_v} & v_0 \\ u_0 & v_0 & 1 \end{pmatrix} \tag{6.7}$$

is a symmetric positive matrix in one-to-one correspondence with the A-matrix since we have:

$$\alpha_v = \sqrt{b_v - v_0^2} \quad ; \quad \gamma = \frac{b_c - u_0\,v_0}{\alpha_v} \quad ; \quad \alpha_u = \sqrt{b_u - u_0^2 - \gamma^2} \tag{6.8}$$

so that we can recover the five components of the A-matrix $(u_0, v_0, \alpha_u, \alpha_v, \gamma)$ from the five parameters of the K-matrix $(u_0, v_0, b_u, b_v, b_c)$, without ambiguity because $\alpha_u > 0$ and $\alpha_v > 0$. Moreover, even if K_i or A_i is known up to a scale factor only, this scale factor is always determined by that fact that the last component of these matrices is always 1. Therefore, we have a one-to-one correspondence between these two matrices, even if defined up to a scale factor.

This K-matrix defines, in fact, the retinal image of the absolute conic at infinity which equation is $0 = m^{\mathrm{T}} \cdot K^{-1} \cdot m$, that is the angular relations between each optical ray, or in other words, the intrinsic calibration or the Euclidean geometry of the system [6.17].

Then, if we know the matrices K_i and K_j we can calculate A_i and A_j, then $R_{ij} = A_i^{-1} \cdot Q_{ij} \cdot A_j$ and $t_{ij} = A_i^{-1} \cdot s_{ij}$. We thus obtain intrinsic and extrinsic (motion) parameters.

But the matrices K_i and K_j depend linearly on Q_{ij}. More precisely (6.6) provides 5 equations on the intrinsic parameters, because it involves an equality between 3x3 symmetric matrices, thus yielding 6 equations, but up to a scale factor. Usually Q is known up to a scale factor, but K is also defined up to a scale factor; therefore this indetermination does not affect the equation.

If we consider the system of equations, for $N + 1$ views, because for any i, j, k, we have

$$Q_{ij} \cdot K_j \cdot Q_{ij}^{\mathrm{T}} = K_i \; ; \; Q_{jk} \cdot K_k \cdot Q_{jk}^{\mathrm{T}} = K_j \; \Rightarrow \; Q_{ki} \cdot K_i \cdot Q_{ki}^{\mathrm{T}} = K_k \; ; \tag{6.9}$$

we obtain, if generic at most 10 linear equations in the 15 unknowns. By an obvious induction, it is clear that, for $N + 1$ views, we will generate at most $5\,N$ equations while we have $5\,(N + 1)$ unknowns in the general case. We thus are left with 5 unknowns in coherence with the previous paragraph.

However, knowing the Qs-representation, we can calculate the calibration parameters (K-matrices which are in one to one correspondence with A-matrices) in several situations:

(1) *If the calibration parameters in one view – say view 0 – are known, we can compute, all other calibration parameters.* Using (6.9) and knowing K_0, we can calculate, K_1 from Q_{10}, K_2 from Q_{20}, etc. and then compute all calibration parameters.

(2) *If 5 additional generic constraints exist between the calibration parameters, we can estimate the calibration parameters in the general case.* Moreover the algebraic complexity of the problem is very low, due to the important fact that all equations are linear, as visible in (6.9).

Such constraints can be stated considering information about the physical sensor: for standard CCD sensors [6.18], for instance, we can assume that $\gamma = 0$ and/or that α_u/α_v is constant. Moreover, variations of some calibration parameters can be predicted during variation of zoom or focus [6.19, 20]. In practice, these equations are to be solved using a statistical framework and a model of the evolution of the parameters, but this is out of the scope of this book.

In any cases, it appears that we can recover the intrinsic parameters with the Qs-representation in an apparently simpler way, than using the projective representation as usually proposed [6.5, 17]. Where does it come from? A very deep reason. As explained before, the Q-matrix corresponds to the collineation of the plane at infinity. From the section on the planar case, we see that if we know the collineation, we have the equation of this plane. But knowing the plane at infinity, means knowing the *affine geometry* of the scene [6.21]. In other words, *the Qs-representation is an intermediate representation of the motion in which the affine geometry of the scene has been made explicit.* A projective representation would have involved F-matrices only, as defined in the next section. An Euclidean representation would have involved rotation matrices, translation vectors and calibration parameters. An affine representation, now, involves Q matrices and s-vectors.

Let us finally explicit a particular case.

(3) *We can recover constant calibration parameters as soon as three views are given, and the rotations between two consecutive views are not performed around the same 3D axis.*

Let us show this statement.

In that case $K_i = K_j = K_k = K$ while $\det(Q_{ij}) = 1$. The equation $Q_{ij} \cdot K \cdot Q_{ij}^{\mathrm{T}} = K$ provides 5 linear equations in the five unknowns $(u_0, v_0, b_u, b_v, b_c)$. In fact, this equation does not allow us to recover these five parameters. Let us explain why. We represent the rotations using the exponential of a skew-symmetric matrix, that is

$$R_{ij} = \mathrm{e}^{\tilde{r}_{ij}}$$

where r is a vector aligned with the rotation axis and which magnitude is equal to the angle of the rotation [6.16].

We can write

$$Q_{ij} = A \cdot R_{ij} \cdot A^{-1} = A \cdot e^{\tilde{r}_{ij}} \cdot A^{-1} = e^{A \cdot \tilde{r}_{ij} \cdot A^{-1}} = e^{K \cdot \tilde{\rho}_{ij}}$$

with $\rho_{ij} = (A \cdot r_{ij})/\det(A)$. This relation is only valid if $K_i = K_j = K$. This relation allows to compute $L_{ij} = K \cdot \tilde{\rho}_{ij}$ as the logarithm matrix of Q_{ij} considering only the unique real solution, since $L_{ij} = K \cdot \tilde{\rho}_{ij}$ is a real matrix. Moreover, it is clear that we cannot calculate all components of K but only in a direction orthogonal to ρ_{ij}. Therefore *we must perform at least two rotations around two non-parallel axes to recover the intrinsic parameters.* Reciprocally, with two such rotations, if we can compute the corresponding Q_{ij} matrices, and thus the corresponding L_{ij} matrices, we can recover K from the two equations $L_{ij} = K \cdot \tilde{\rho}_{ij}$. This is coherent to what has been observed by Viéville [6.8].

We thus have not only clarified that (1) the "5 missing parameters" correspond to the parameters of the absolute conic in the plane at infinity (a 2D conic is defined by 5 parameters), but (2) have shown that there are at least two situations for which we can calculate them, if the initial intrinsic calibration parameters are known, or if they the intrinsic calibration parameters are constant. However, (3) it is clear but we cannot recover these parameters without at least 5 constraints.

Finally, let us point out that it is possible for a rather large class of real scenes, to infer the collineation of the plane at infinity, i.e. the Q-matrix, considering points "at the horizon", that is with a negligible depth [6.21]. This segmentation process, although not rigorous is quite efficient and is discussed elsewhere [6.22].

Structure from Motion Using the Qs-Representation. In order to increase our understanding of the potentialities of this representation, let us explicit a very useful, but oversimple result. The formula is obtained from (6.2) very easily:

Proposition 6.1. Using the Qs-representation, the 3D affine depth of an image point can be directly recovered, without an explicit knowledge of the intrinsic calibration (which can change at any time) and of the Euclidean displacements, since

$$Z_j \ (Q_{ij} \cdot m_j \wedge m_i) + (s_{ij} \wedge m_i) = 0 \ .$$

This result means that we can directly build a depth map, the Q-matrix and s-vector being known.

In fact, we obtain only *one equation* because this equation vanishes in the direction of m_i and, considering (6.11) in the direction of s_{ij} also.

Moreover, we can relate the depth from one frame to another since we obtain very easily

$$Z_j \ (s_{ij} \wedge m_j) = Z_i \ (\tilde{s}_{ij} \cdot Q_{ij}) \cdot m_i \ . \tag{6.10}$$

As expected the depths Z_j are related to the depths Z_i up to common expansion factor since the scale of the Qs-representation is not known.

This result is very useful in practice for the following reason: if we are interested in building a depth map, we can deal with this Qs-representation, without recovering the Euclidean parameters of the system, which is generally not possible as discussed previously.

However, one must understand that we do not recover the Euclidean structure of the scene, since the depth Z_j is a function of m_j and not of M_j. We must recover the A-matrix, i.e. the intrinsic calibration parameters, since $m_j = \lambda A \cdot M_j$, to reconstruct the Euclidean depth map. If not, we only reconstruct the depth map, up to an *affine transformation* of the image, given by the matrix A, and this corresponds to an affine transformation of the 3D space.

More generally, considering (6.3) and following the same method as in [6.23], when calibration was taken into account, we can make the following count of equations. Given $N + 1$ views of a monocular image sequence and P points, the problem of recovering the depths up to a scale factor can be solved only if we have more equations than unknowns, more precisely:

$$\underbrace{(1 + N)\,P\,(\text{depth}) + 11\,N - 1\,(QS\text{-parameters})}_{\text{unknowns}} -3\,N\,P\,(\text{equations}) \leq 0$$

$$\Leftrightarrow \quad P \geq \frac{11\,N - 1}{2\,N - 1}.$$

The problem is solved up to a finite number of possibilities if we have the same number of equations as unknowns, if we have more equations than unknowns, and if these equations corresponds to a physical non-degenerated solution, this solution is in general unique.

Let us explicit the minimum number $P_{\min}$ of points to solve (6.3):

$N + 1$	3	4	5	6	...	∞
$P_{\min}$	7	$7 - \frac{3}{5}$	$7 - \frac{6}{7}$	6	...	$6 - \frac{1}{2}$

One application of this method is that for $P > P_{\min}$ there exists some internal relations between the coordinates of the points, neither dependent on motion nor on depth, thus **invariant** [6.24]. It is known that given more than 8 points in 2 views, there exists invariants [6.9]. The previous count allows us to conjecture that given more than 7 points in 3, 4 or 5 views or more than 6 points in 6 views or more, there exists invariants. These invariants, are related to the rigidity constraint, i.e. they are verified if and only if the points belong to the same rigid object.

6.2.3 From Affine to a Projective Parameterization

Decomposition of the Qs-Representation. If we eliminate Z_i and Z_j in (6.3) we obtain

$$m_i^T \underbrace{(\tilde{s}_{ij} \cdot Q_{ij})}_{F_{ij}} \cdot m_j = 0 \; . \tag{6.11}$$

The matrix $F_{ij} = \tilde{s}_{ij} \cdot Q_{ij}$ is the *fundamental matrix* and is also called the "essential matrix in the uncalibrated case" considering the original Longuet–Higgins equation [6.1], now generalized.

In fact, if we consider that the only information available is related to the retinal correspondences between points, without any knowledge about the depths Z_i, (6.11) is the only equation that can be derived, as known from a long time in the calibrated case [6.1] and recently confirmed in the uncalibrated case [6.17, 25]. However, note that, in an image sequence, such equations are available between each pair of consecutive or non-consecutive frames, although all F-matrices are not expected to be independent. We are going to clarify this problem now.

There is also a deep relationship between the Qs-representation and the fundamental matrix defined by [6.17, 25], but if we want to understand this relationship, we need the following decomposition of a Q-matrix, given the corresponding s-vector:

$$Q_{ij} = S_{ij} + s_{ij} \cdot r_{ij}^T \quad \text{with} \quad s_{ij}^T S_{ij} = 0 \; ;$$
$$\Leftrightarrow r_{ij} = \frac{Q_{ij}^T \cdot s_{ij}}{||s_{ij}||^2} \quad ; \quad S_{ij} = -\frac{\tilde{s}_{ij}^2}{||s_{ij}||^2} \cdot Q_{ij} \; . \tag{6.12}$$

In this decomposition, the image of the s-vector through the transpose of the Q-matrix has been made explicit. See [6.26] for a geometric interpretation. In other words, the transpose of the Q-matrix is decomposed into its restriction to the vectorial line generated by s_{ij} and its restriction to $s_{ij}^\perp$, represented by S_{ij}. The reader can easily verify, from (6.12), that this decomposition is one-to-one.

There is also a one-to-one correspondence between F-matrices and S-matrices since we have

$$F_{ij} = \tilde{s}_{ij} \cdot S_{ij} \quad ; \quad s_{ij}^T \cdot S_{ij} = 0 \; ;$$
$$\Leftrightarrow S_{ij} = -\frac{\tilde{s}_{ij}}{||s_{ij}||^2} \cdot F_{ij} \quad ; \quad s_{ij}^T \cdot F_{ij} = 0 \; . \tag{6.13}$$

Moreover, although in this equation, both quantities are defined only up to scale factor, if we use the following two constraints: $||s_{ij}||^2 = 1$ and $||S_{ij}||^2 = 1$, in addition to the 3 constraints given by $s_{ij}^T \cdot S_{ij} = 0$, we have a $3 - 1$ (s-vector) $+ 9 - 3 - 1$ (S-matrix) $= 7$ (parameters) representation of F_{ij}, as expected [6.25]: a F-matrix is defined by 7 parameters. (The matrix has 9 coefficients but is defined only up to 1 scale factor and is subject to 1 constraint $\det(F_{ij}) = 0$). However, in our case we have entirely defined the

F-matrix using (6.11). A local parameterization of the F-matrix based on these relations will be proposed in a next section.

If only two views are given there is no way to recover the Q-matrix from point correspondences, since only S_{ij} and s_{ij} are calculate from a F-matrix, whereas r_{ij} is not. But considering S-matrices and s-vectors in an image sequence, we can relate all r-vectors by the following crucial relation, obtained by making explicit the r-vectors in the formula $Q_{ij} = Q_{ik} \cdot Q_{kj}$ and using some useful relations between this elements, given in (6.21):

$$r_{ij} = r_{kj} + q_{ijk} \, ;$$

$$q_{ijk} = \frac{(s_{ik}^T \cdot s_{ik}) \, S_{kj}^T \cdot S_{ik}^T \cdot s_{ij} + (s_{ij}^T \cdot s_{ik}) \, S_{ij}^T \cdot s_{ik}}{(s_{ij}^T \cdot s_{ij}) \, (s_{ik}^T \cdot s_{ik}) - (s_{ij}^T \cdot s_{ik})^2} \, . \tag{6.14}$$

Please note that this expression is coherent with respect to the different scale factors indeterminations as follows. If $Q_{ij} = S_{ij} + s_{ij} \cdot r_{ij}^T$ is defined up to a scale factor λ_{ij} and s_{ij} is defined up to a scale factor μ_{ij}, as discussed previously, S_{ij} is defined up to λ_{ij}, r_{ij} is defined up to λ_{ij}/μ_{ij} and F_{ij} is defined up to $\lambda_{ij} \, \mu_{ij}$. Introducing these scale factors in (6.14) it appears that q_{ijk} is defined up to a scale factor λ_{ij}/μ_{ij} like r_{ij}. Moreover it is clear that we these scales are not independent because we must have $\mu_{ij} = (\lambda_{ij}/\lambda_{kj}) \, \mu_{kj}$ in coherence with the discussion of the previous paragraph.

We also can compute all r_{ji} from r_{ij}, if we explicit r-vectors in the relation $Q_{ij} \cdot Q_{ji} = I$, if we have

$$r_{ji} = S_{ji}^T \cdot r_{ij} - \frac{s_{ij}}{s_{ij}^T \cdot s_{ij}} \, . \tag{6.15}$$

From these two relations, we can – know r_{kj} – recover r_{ij}, thus Q_{ij} and Q_{kj} and Q_{ik}, and from frame to frame all Q-matrices. Finally, we have established that all r-vectors are entirely determined from the knowledge of the S-matrices or equivalently the F-matrices, up to one of them, say r_{01}. This means that all F-matrices defines all Qs-representations, except 1 r-vector, so define at least 11 $N - 4$ parameters, and at most 11 $N - 1$ parameters.

Can we recover the 3 last parameters corresponding to the ultimate r-vector? The answer is no, and this can be easily shown by induction: let us assume we have recovered all r-vectors considering $N+1$ views and S-matrices and s-vectors, having 11 $N - 1$ parameters in this representation. Therefore, considering the N-first views we also must have been able to recover all r-vectors since we have at least 11 parameters less, in this case (a Q-matrix and a s-vector, up to a scale factor are undefined). By induction, for two views, i.e. $N = 1$, we must have still 11 $N - 1 = 11$ parameters, which is not the case, a F-matrix being a function of 7 parameters, i.e. $\{11 \, N - 4\}_{N+1=2}$, only. Therefore, 3 parameters, are indeed "missing".

The Projective Representation of Motion in an Image Sequence. It has been made explicit in the previous paragraph that we can obtain S-matrices and s-vectors between consecutive pair of frames and q-vectors between consecutive triplets of frames, considering only retinal correspondences, i.e. estimating only F-matrices.

It appears that the inverse is true and, in addition to that, *we can obtain a minimal parameterization of the retinal correspondences, using this Ssq-representation.*

Let us develop this point now.

Considering $N + 1$ views, as shown in Fig. 6.4, and only retinal point correspondences, i.e. assuming that only F-matrices can be estimated, we parameterize the uncalibrated motion, as follows:

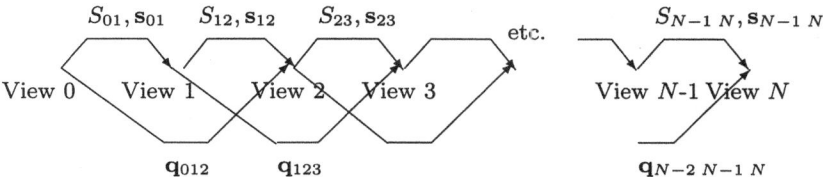

Fig. 6.4. Using S-matrices, s-vectors and q-vectors in an image sequence

(1) S-matrices and s-vectors between the N consecutive pairs of views:
   $\{S_{i-1,i}, s_{i-1,i}\}_{i=1...N}$.
(2) q-vectors between the $N - 1$ consecutive triplets of views:
   $\{q_{i-1,i,i+1}\}_{i=1...N-1}$.

We use 12 numbers for S-matrices and s-vectors (a 3×3 matrix and 3×1 vector) but subject to 3 constraints through the equation $s_{ij}^{\mathrm{T}} \cdot S_{ij} = 0$, i.e. 9 N parameters. We also use use 3 $(N - 1)$ additional parameters, for the q-vectors. Moreover, as visible in (6.37), each pair of S-matrix and s-vector is subject to a scale indetermination (i.e. N parameters less), while there is another global scale indetermination in (6.37), when considering the quantities $(\|S\|/\|s\|, \|q\|)$, as detailed in the previous paragraph.

These quantities are therefore defined by 9 $N+3$ $(N-1)-N-1 = 11$ $N-4$ parameters.

Now this representation is sufficient since:

− All S-matrices and s-vectors can be generated from $\{S_{i-1,i}, s_{i-1,i}\}_{i=1...N}$ (i.e. S-matrices and s-vectors between consecutive frames), through the relations, easily obtained from (6.21):

$$
\left\{
\begin{aligned}
0 &= S_{ij} \cdot s_{ji} , \\
S_{ji} &= -\det(Q_{ij}) \, \frac{\tilde{s}_{ji}}{||s_{ji}||^2} \cdot S_{ij}^{\mathrm{T}} \cdot \tilde{s}_{ij} , \\
s_{ik} &= S_{ij} \cdot s_{jk} + (q_{ijk}^{\mathrm{T}} \cdot s_{jk}) \, s_{ij} , \\
S_{ik} &= \frac{\tilde{s}_{ik}^2}{||s_{ik}||^2} \cdot \left[S_{ij} \cdot S_{jk} \right. \\
&\quad \left. + s_{ij} \cdot \left(S_{jk}^{\mathrm{T}} \cdot q_{ijk} + \frac{s_{kj}}{||s_{kj}||^2} \right)^{\mathrm{T}} \right] .
\end{aligned}
\right. \tag{6.16}
$$

- All F-matrices are in one to one correspondence with S-matrices, as shown in (6.13).

Reciprocally, given all F-matrices in an image sequence, we can calculate all S-matrices and s-vectors, using (6.13) and the relation $s_{ij}^{\mathrm{T}} \cdot S_{ij} = 0$, and all q-vectors using (6.14).

With this parametrization, again, considering the point location in the first two views, we can recover its location with the $N - 1$ other views.

We will call this representation the Ssq-representation in the sequel.

Comparing to the previous representation, 3 parameters are missing. This comes from the fact that we have not assume to have a knowledge of the plane at infinity defined between the view i and the view j by r_{ij}.

Let us now study how to relate this representation to affine and Euclidean parameters, i.e. K-matrices and r-vectors, in the present case.

Relation Between the Ssq-Representation and Euclidean or Affine Parameters. If we assume that we know the F-matrices, i.e. S-matrices and s-vectors, we have to recover r-vectors and K-matrices. This problem can be analysed as follows.

Considering (6.9) in which we explicit the S-matrix, r-vector and s-vector, we obtain, after some simplifications, the following set of equations

$$
\left\{
\begin{aligned}
S_{ij} \cdot K_j \cdot S_{ij}^{\mathrm{T}} &= \left(I - \frac{s_{ij} \cdot s_{ij}^{\mathrm{T}}}{s_{ij}^{\mathrm{T}} \cdot s_{ij}} \right) \cdot K_i \cdot \left(I - \frac{s_{ij} \cdot s_{ij}^{\mathrm{T}}}{s_{ij}^{\mathrm{T}} \cdot s_{ij}} \right) , \\
S_{ij} \cdot K_j \cdot r_{ij} &= \left(I - \frac{s_{ij} \cdot s_{ij}^{\mathrm{T}}}{s_{ij}^{\mathrm{T}} \cdot s_{ij}} \right) \cdot K_i \cdot \frac{s_{ij}}{s_{ij}^{\mathrm{T}} \cdot s_{ij}}
\end{aligned}
\right. \tag{6.17}
$$

since we have $r_{ij}^{\mathrm{T}} \cdot K_j \cdot r_{ij} = s_{ij}^{\mathrm{T}} \cdot K_i \cdot s_{ij}$.

The first equation provides 9 linear homogeneous equations for the K-matrices, but not all independent. Because the matrices are symmetric, we have 6 equations only, but this matrix expression is null in the direction of s_{ij}, so 3 equations are always verified. As a consequence it provides 3 linear homogeneous equations only.

It is true that S_{ij} is known up to a scale factor only but K_j also, and this scale factor could be integrated in the definition of K_j since this matrix is only defined up to a scale factor, and the equation is coherent.

Finally, it is easy to remark that the matrix K_i cannot be recovered in the direction of s_{ij} since if we write: $K_i = \kappa_i\, s_{ij} \cdot s_{ij}^{\mathrm{T}} + \Lambda_i$ the quantity κ_i disappears in the equation. Similarly, K_j cannot be recovered in the direction of s_{ji}.

The second equation provides, if r_{ij} is known, an additional set of 2 linear homogeneous equations for the K-matrices, because this vectorial expression is null in the direction of s_{ij}. The scale factors are coherent with respect to the first equation. If generic these $3 + 2 = 5$ equations are independent. We obtain in fact the 5 equations (6.9) again.

If r_{ij} is not known or partially known, but K_j has been estimated, the second equation provides a set of 2 linear homogeneous equations for this quantity. We have only 2 equations since this vectorial expression is null in the direction of s_{ij}. Moreover, r_{ij} can not be recovered in the direction $K_j^{-1} \cdot s_{ji}$. We only have homogeneous equations since the K-matrix and S-matrix are defined up to a scale factor.

Let us now apply these equations when the 3 motions between 3 views is given. From the first equation of (6.17), using S_{01}, S_{21}, s_{01} and s_{21}, we obtain 6 equations to determine the K-matrices. In particular, if $K_0 = K_1 = K_2 = K$ the problem can be solved. More generally, having a model on the evolution of the intrinsic parameters, i.e. of the K-matrices one can combine it with this 6 measurement equations.

From the second equation of (6.17), using the estimation of K_0, K_2 and K_1, we recover r_{01} up to a scale factor and not in the direction $s'_{01} = K_0^{-1} \cdot s_{01}$, i.e. a vector q_0 such that $\lambda_0\, q_0 = r_{01} + \mu_0 s'_{01}$. Similarly, we recover a vector q_2 such that

$$\lambda_2\, q_2 = r_{21} + \mu_2 \underbrace{K_2^{-1} \cdot s_{21}}_{s'_{21}} \;.$$

But we also have recovered $q_{012} = \lambda_1\, q_1$ up to a scale factor, i.e. we can write $\lambda_1\, q_1 = r_{01} - r_{21}$ for some λ_1. Using the relation $r_{ij} \cdot s_{ji} = -1$ we can eliminate these scale factors and finally obtain

$$
\begin{cases}
r_{01} = -\dfrac{s'_{01}}{s'^{T}_{01} \cdot s'_{10}} - \dfrac{q_0}{q_0^{T} \cdot s'_{10}} \\[2ex]
\qquad - \dfrac{\left| q_1, \; \dfrac{s'_{21}}{s'^{T}_{21} \cdot s'_{12}} - \dfrac{s'_{01}}{s'^{T}_{01} \cdot s'_{10}}, \; \dfrac{q_2}{q_2^{T} \cdot s'_{12}} - \dfrac{s'_{21}}{s_{21}^{T} \cdot s'_{12}} \right|}{\left| q_1, \; \dfrac{s'_{01}}{s^{T}_{01} \cdot s'_{10}} - \dfrac{q_0}{q_0^{T} \cdot s'_{10}}, \; \dfrac{q_2}{q_2^{T} \cdot s'_{12}} - \dfrac{s'_{21}}{s_{21}^{T} \cdot s'_{12}} \right|} \dfrac{s_{01}}{s'^{T}_{01} \cdot s'_{10}} \\[4ex]
r_{21} = -\dfrac{s_{21}}{s'^{T}_{21} \cdot s'_{12}} - \dfrac{q_2}{q_2^{T} \cdot s'_{12}} \\[2ex]
\qquad - \dfrac{\left| q_1, \; \dfrac{s'_{01}}{s'^{T}_{01} \cdot s'_{10}} - \dfrac{q_0}{q_0^{T} \cdot s'_{10}}, \; \dfrac{s'_{21}}{s_{21}^{T} \cdot s'_{12}} - \dfrac{s'_{01}}{s'^{T}_{01} \cdot s'_{10}} \right|}{\left| q_1, \; \dfrac{s'_{01}}{s'^{T}_{01} \cdot s'_{10}} - \dfrac{q_0}{q_0^{T} \cdot s'_{10}}, \; \dfrac{q_2}{q_2^{T} \cdot s'_{12}} - \dfrac{s'_{21}}{s_{21}^{T} \cdot s'_{12}} \right|} \dfrac{s_{21}}{s'^{T}_{21} \cdot s'_{12}}
\end{cases}
\qquad \text{(6.18)}
$$

We thus can recover the r-vectors from the knowledge of the K-matrices, without an explicit recovery the Euclidean motion and the calibration parameters.

Since we also have recovered S and s up to a *common* scale factor, we have to entirely recover the Q-matrix up to a scale factor. Unfortunately, it appears that we must compute the Euclidean calibration (i.e. the K-matrix) to obtain the r vectors.

Unfortunately again, this process is feasible if and only if, either the K-matrices are constant, or some other hypotheses are available on the calibration of the system.

Remarks on the Algebraic Complexity of the Projective Parameterization. It might be surprising that we have spent quite a lot of efforts to construct a rather complex Ssq-representation of the $11\,N - 4$ parameters, in the projective case, whereas we could have "directly used the F-matrices". This other alternative is – in fact – far from being obvious and we would like to explain why now.

As for the Qs-representation, using (6.4), we have the following relations between F-matrices in an image sequence:

$$
\begin{aligned}
F_{ii} &= 0 \,, \\
F_{ij} &= \det(Q_{ij})\, F_{ji}^{\mathrm{T}} \,, \\
F_{ij} &= \det(Q_{ik})\, \left(Q_{ki}^{\mathrm{T}} \cdot F_{kj} + F_{ki}^{\mathrm{T}} \cdot Q_{kj} \right) \\
&= \det(Q_{ik})\, \left[Q_{ki}^{\mathrm{T}} \cdot (\tilde{s}_{kj} - \tilde{s}_{ki}) \cdot Q_{kj} \right] \\
&= \det(Q_{ik})\, \left[\frac{\tilde{s}_{ik}}{\det(Q_{ik})} \cdot Q_{ij} - Q_{ji}^{\mathrm{T}} \cdot \frac{\tilde{s}_{jk}}{\det(Q_{jk})} \right]
\end{aligned}
\tag{6.19}
$$

from which we deduce the fundamental relation

$$
s_{ik}^{\mathrm{T}} \cdot F_{ij} \cdot s_{jk} = 0 \qquad \Leftrightarrow \qquad s_{jk}^{\mathrm{T}} \cdot F_{ji} \cdot s_{ik} = 0
\tag{6.20}
$$

since $s_{ij} \wedge s_{ik} = F_{ij} \cdot s_{jk}$.

However, it is not possible to estimate a F-matrix from other F-matrices without using a Q-matrix, which does not allow to base our representation on F-matrices only.

Moreover, if we consider $N + 1$ views we can define $\frac{1}{2} N\,(N + 1)$ F-matrices and $N\,(N + 1)$ corresponding epipoles. However, considering (6.20) for any indices, we generate $\frac{1}{2} N\,(N - 1)\,(N - 2)$ constraints, obviously not all independent.

As a consequence, we must not only compute F-matrices between consecutive frames (i.e. F_{01}, F_{12}, F_{23}, etc.) which would provide no more than $7\,N$ parameters, but also F-matrices between non-consecutive frames (i.e. F_{02}, F_{03}, F_{13}, etc.), although very complex algebraic relations between these matrices might occur.

This is why we have introduced this new representation, using r-vectors, and S-matrices. The basic idea was thus to *decompose* the F-matrices and

Q-matrices in order to facilitate the identification of the implicit relations. Moreover, this algebraic construction is a very rich structure since we have, for instance:

$$r_{ij}^{\mathrm{T}} \cdot s_{ji} = -1 \, ,$$

$$r_{ij}^{\mathrm{T}} \cdot s_{ij} = -\frac{s_{ij}^{\mathrm{T}} \cdot s_{ji}}{s_{ji}^{\mathrm{T}} \cdot s_{ji}} \, ,$$

$$r_{ij}^{\mathrm{T}} s_{jk} = \frac{s_{ij}^{\mathrm{T}} \cdot s_{ik}}{s_{ij}^{\mathrm{T}} \cdot s_{ij}} - 1 \, ,$$

$$I = S_{ij} \cdot S_{ji} + \frac{s_{ij} \cdot s_{ij}^{\mathrm{T}}}{s_{ij}^{\mathrm{T}} \cdot s_{ij}} \, , \qquad (6.21)$$

$$\left(I - \frac{s_{ik} \cdot s_{ik}^{\mathrm{T}}}{s_{ik}^{\mathrm{T}} \cdot s_{ik}} \right) \cdot s_{ij} = S_{ik} \cdot s_{kj} \, ,$$

$$r_{kj} = S_{kj}^{\mathrm{T}} \cdot r_{ik} - \frac{(s_{ij}^{\mathrm{T}} \cdot s_{ik}) \, S_{kj}^{\mathrm{T}} \cdot S_{ik}^{\mathrm{T}} \cdot s_{ij} - (s_{ij}^{\mathrm{T}} \cdot s_{ij}) \, S_{ij}^{\mathrm{T}} \cdot s_{ik}}{(s_{ij}^{\mathrm{T}} \cdot s_{ij}) \, (s_{ik}^{\mathrm{T}} \cdot s_{ik}) - (s_{ij}^{\mathrm{T}} \cdot s_{ik})^2} \, ,$$

$$S_{ij} = S_{ik} \cdot S_{kj} + s_{ij} \cdot (r_{kj} - r_{ij})^{\mathrm{T}} + s_{ik} \cdot (S_{kj}^{\mathrm{T}} \cdot r_{ik} - r_{kj})^{\mathrm{T}} \, ;$$

all these relations being easily obtained introducing the S-matrix and s-vector in the relations of (6.4). They are used for some technical developments, in this book. In particular they allow the reader to easily derive (6.14), (6.16) and (6.17).

Structure from Motion Using the Ssq-Representation. If we do not known the Q-matrix but only the related S-matrix, as when considering F-matrices, we obtain:

Proposition 6.2. When the r-vector of the Qs-representation is not known, the 3D depth of an image point can be directly recovered, up to a collineation of the depth map, without an explicit knowledge of the intrinsic calibration (which can change at any time) and of the Euclidean displacements, since

$$\underbrace{\frac{Z_j}{1 + Z_j \, (r_{ij}^{\mathrm{T}} \cdot m_j)}}_{\zeta_j} (S_{ij} \cdot m_j \wedge m_i) + (s_{ij} \wedge m_i) = 0 \, .$$

The quantity ζ_j is related to the Euclidean depth through a special collineation. The corresponding coefficients are functions of the retinal location. In such a case we reconstruct the depth map, not only up to affine transformation of the retinal plane, but also up to projective transform of the depth. These transformations have been already made explicit [6.27], in the case of a stereo rig.

Therefore, we can design the following strategy: compute a *precise* 3D reconstruction, up to an affine/projective transformation, *without calibration,* and then estimate the previous transformation from the knowledge of the calibration parameters.

As in the affine case, the depth from one frame to another can be predicted up to an unknown expansion factor, using (6.10). Note that this equation is only function of s_{ij} and $F_{ij} = \tilde{s}_{ij} \cdot Q_{ij}$ and thus depends on projective quantities only.

6.2.4 Conclusion: Choosing a Composite Representation

We have obtained the following hierarchy of parameters:

Projective Motion Computation	P $= 11\,N - 4$ for $N + 1$ views
Affine Motion Computation	A $=$ P $+ 3$ parameters
Euclidean Motion Computation	M $=$ A $+ 5$ parameters

The 3 and 5 "missing" parameters in the affine and Euclidean representation have the following geometrical interpretation:

The *affine calibration* corresponds to the identification of the plane at infinity. A plane is defined by 3 parameters, as expected. This is represented by the r-vector in our equations. As soon as this quantity is known in one view, it can be predicted in the whole image sequence.

The *Euclidean calibration* corresponds to the identification of the image of the absolute conic in the plane at infinity. The planar equation of a conic is defined by 5 parameters, as expected. This is represented by the K-matrix in our equations. As soon as this quantity is known in one view, it can be predicted in the whole image sequence. Knowing the K-matrix the r-vector can be computed, we thus do not have $5 + 3$ independent parameters.

Now, considering these three representations, it is clear that the Qs-*representation* is much simpler because: It only requires unconstrained matrices and vectors between consecutive frames. It is directly related to the 3D depth of the points in a retinal frame of reference.

In the projective case, it has only one indetermination of 3 parameters related to the absence of knowledge about 1 r-vector between one of the image pair. But this can be easily avoided. In fact we propose now another *composite* representation of the $11\,N - 4$ parameters, builded as follows.

We consider: (a) The F-matrix defined between view 0 and 1. (b) Qs-representations between other pairs of consecutive frames. (c) The relative scale factor between the F-matrix and other Q-matrices.

We obtain a representation of $7 + 11\ (N - 1) - 1 + 1$, that is $11\ N - 1$ parameters, as expected. This is strictly equivalent to the representation proposed in Fig. 6.4, because we have established a one-to-one correspondence between Q-matrices and q-vectors, s-vectors and S-matrices, except in the first pair of frames, where only the S-matrix, or equivalently the F-matrix, is measurable.

Note that each point correspondences bring one equation for the first couple of views and two equations for each additional view, i.e. $2\ N - 1$ equations along the image sequence. Let us consider a sequence of $N + 1$ views and Pt correspondences between points, in order to solve the system, we must have

$$[2\ N - 1]\ Pt \le 11\ N - 4\,,$$

therefore at least 7 points for 2 views as already known [6.25], and 6 points for 3 views as in [6.28], but also 6 points for more than 3 views. Thus, we need always more than 5 point correspondences.

This representation is schematized in Fig. 6.5.

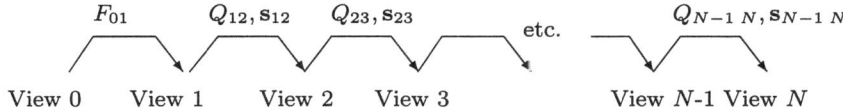

Fig. 6.5. Using a composite representation for the retinal motion in an image sequence

This final representation will be implemented and experimented in the last section.

6.3 Representations of the Retinal Motion for Points and Lines

Let us now introduce lines as tokens and try to combine points and lines in our equations.

6.3.1 The Retinal Motion for Lines

Considering Lines as Tokens. Lines are represented here by their Plücker coordinates as follows. Let D be a 3D line. We consider the unary vector $\boldsymbol{L}_i$, with $\|\boldsymbol{L}_i\| = 1$, which is aligned with the 3D line direction, and the vector

$$\boldsymbol{N}_i = \boldsymbol{C}_i \boldsymbol{M} \wedge \boldsymbol{L}_i \tag{6.22}$$

for any point $M \in D$, as illustrated in Fig. 6.6. It is straightforward to verify that $\boldsymbol{N}_i$ is orthogonal to $\boldsymbol{L}_i$ and does not depend on M [6.16]. Moreover,

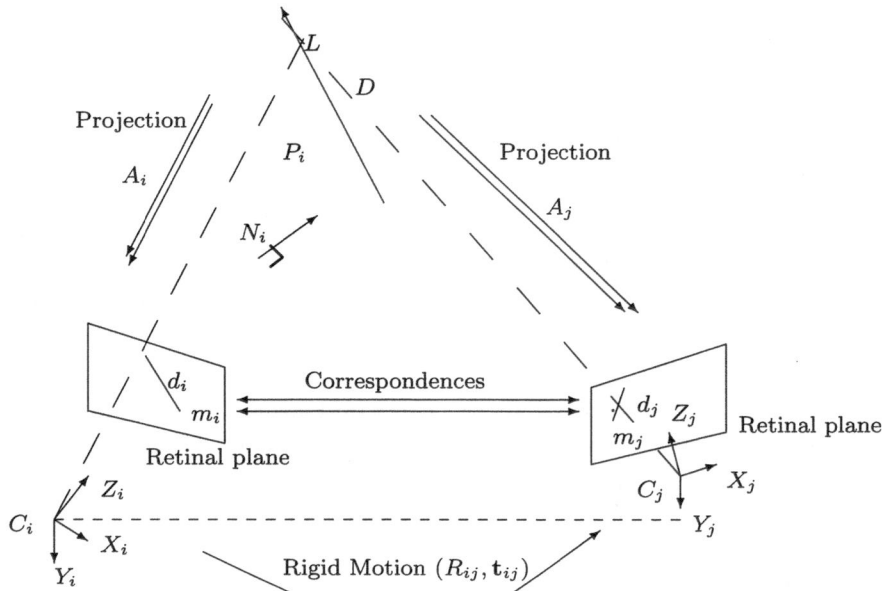

Fig. 6.6. Elements used in the definition of lines in motion

the magnitude of N_i, $\|N_i\|$, is equal to the distance from D to the origin C_i. Then, 3D lines are represented by two 3D vectors (L_i, N_i) with the two quadratic constraints $L_i^T \cdot L_i = 1$ and $L_i^T \cdot N_i = 0$, yielding a parameter space of dimension 4, as expected.

Given two points $M^a \in D$ and $M^b \in D$ with $M^a \neq M^b$ we have, up to an $\epsilon = \pm 1$

$$\begin{cases} L_i &= \epsilon \dfrac{M^b - M^a}{\|M^b - M^a\|} \\[2mm] N_i &= \epsilon \dfrac{M^a \wedge M^b}{\|M^b - M^a\|} \end{cases} \tag{6.23}$$

as the reader can easily verify.

A 3D line D projects onto a line d_i in the camera indexed by i. Let P_i be the plane containing D, the origin C_i and the projection d_i. Obviously N_i is normal to this plane.

We write the equation of a 2D line d_i, projection of a 3D line D, in the camera indexed by i:

$$m \in d_i \quad \Leftrightarrow \quad m^T \cdot n_i = 0 \, ; \tag{6.24}$$

the vector n_i is an homogeneous vector.

Moreover, to the 3D line direction corresponds a point at infinity, i.e a direction in the Euclidean space; let us write l_i the projection of this point at infinity, onto the retinal plane, i.e. the corresponding vanishing point. This homogeneous vector corresponds to the "uncalibrated" information about the 3D line direction.

More precisely, using (6.1) and the previous definitions, we immediately obtain

$$\begin{cases} \boldsymbol{n}_i = \dfrac{1}{\alpha_i} A_i^{-1\,\mathrm{T}} \cdot \boldsymbol{N}_i & \Leftrightarrow & \boldsymbol{N}_i = \alpha_i\, A_i^{\mathrm{T}} \cdot \boldsymbol{n}_i\,, \\[2mm] \boldsymbol{l}_i = \dfrac{1}{\beta_i} A_i \cdot \boldsymbol{L}_i & \Leftrightarrow & \boldsymbol{L}_i = \dfrac{1}{||A_i^{-1} \cdot \boldsymbol{l}_i||}\, A_i^{-1} \cdot \boldsymbol{l}_i\,. \end{cases} \tag{6.25}$$

We have chosen $\alpha_i = ||A_i^{-1\,\mathrm{T}} \cdot \boldsymbol{N}_i||$ and $\beta_i = ||A_i \cdot \boldsymbol{L}_i||$ to have $\boldsymbol{l}_i$ and $\boldsymbol{n}_i$ unary.

In fact, α_i is also the "uncalibrated structure" of the 3D line, since it must be estimated to recover the distance from the line to the origin, i.e. $||\boldsymbol{N}_i||$. On the contrary, β_i is only a normalization factor.

It is clear from these equations that neither $\boldsymbol{L}_i$, nor $\boldsymbol{N}_i$ can be recovered without knowing the intrinsic parameters, i.e. A_i, whereas α_i and $\boldsymbol{l}_i$ might be estimated directly from the retinal correspondences, as it will discussed in the sequel.

If m_i is a point on the 2D line, the parametric representation of the line is very simple: $m_i + \lambda\, \boldsymbol{l}_i$ for $\lambda \in \mathcal{R}$.

Given two points m^a and m^b on the 2D line d_i, projection of two points $M^a \in D$ and $M^b \in D$ we have, from (6.23) and (6.1), up to an $\epsilon = \pm 1$:

$$\begin{cases} \boldsymbol{l}_i = \underbrace{\dfrac{\epsilon}{||M^b - M^a||\, \beta_i}}_{\kappa_i} \left(Z^b\, m^b - Z^a\, m^a \right)\,, \\[4mm] \boldsymbol{n}_i = \underbrace{\dfrac{\epsilon\, Z^a\, Z^b}{||M^b - M^a||\, \alpha_i\, \det(A_i)}}_{\rho_i} \left(m^a \wedge m^b \right) \end{cases} \tag{6.26}$$

as the reader can easily verify again.[3]

Using the Qs-Representation for Lines. If we consider that each point of the line has a rigid motion of the form: $M_i = R_{ij} \cdot M_j + \boldsymbol{t}_{ij}$, and applying this equation to the Plücker representation of a 3D line allows to compute the 3D correspondences, for a line, between two frames. Combining (6.22) and (6.2) yields after a few algebra [6.29]:

$$\boldsymbol{L}_i = R_{ij} \cdot \boldsymbol{L}_j \quad ; \quad \boldsymbol{N}_i = R_{ij} \cdot \boldsymbol{N}_j + \boldsymbol{t}_{ij} \wedge \boldsymbol{L}_i\,. \tag{6.27}$$

This equation defines the 3D correspondence, for a line, between two frames.

---

[3] We make use of the following relation:
$$(M \cdot \boldsymbol{x}) \wedge (M \cdot \boldsymbol{y}) = \underbrace{\det(M)\, M^{-1\,\mathrm{T}}}_{M^*} \cdot (\boldsymbol{x} \wedge \boldsymbol{y})\,.$$

Let us now compute the 2D correspondence between the two projections of a 3D line, in frame i and j. The 2D lines d_i and d_j are represented by their normal vectors n_i and n_j. In fact, the same computation can be performed combining (6.25) with (6.27) rewritten in the reverse form: $N_j = R_{ij}^T \cdot N_i + s_{ji} \wedge L_j$. We thus have, performing a few algebra

$$\alpha_j \, n_j = \alpha_i \, Q_{ij}^T \cdot n_i + \frac{\beta_j}{\det(A_j)} \, s_{ji} \wedge l_j \; . \tag{6.28}$$

Similarly, we have from $L_i = R_{ij} \cdot L_j$

$$\beta_i \, l_i = \beta_j \, Q_{ij} \cdot l_j \; . \tag{6.29}$$

So, in both cases (points and lines), the 2D correspondences between features related to the projection of a rigid object can be easily represented using the projection of the 3D features, and the Qs-representation.

Let us analyze these equations. From (6.28) and (6.25) we obtain

$$l_j = \frac{\left[\underbrace{\|A_j^{-1\,T} \cdot N_j\| \; \det(A_j)}_{\alpha_j} \right]}{\left[\underbrace{\|A_j \cdot L\|}_{\beta_j} (s_{ij}^T \cdot n_i) \right]} \; (Q_{ij}^T \cdot n_i \wedge n_j) \tag{6.30}$$

and replacing in (6.28) yields

$$0 = \left[\alpha_j \, (s_{ji}^T \cdot n_j) + \alpha_i \, (s_{ij}^T \cdot n_i) \right] (Q_{ij}^T \cdot n_i - n_j)$$
$$\Rightarrow \quad \frac{\alpha_i}{\alpha_j} = -\frac{(s_{ji}^T \cdot n_j)}{(s_{ij}^T \cdot n_i)} \; . \tag{6.31}$$

In other words we *only* recover:

(1) l_j up to a scale factor, i.e. the projection of the direction of the 3D line onto the retina, or its vanishing point.
(2) α_i/α_j, i.e. a quantity in one-to-one correspondence with the distance from the line to the origin. There is thus a scale factor indetermination in this case, which corresponds to the monocular paradigm.

Affine Reconstruction of Lines. Can we generalize Prop. 6.1 considering lines? Yes. But this is not obvious. Let us explain how.

It is easily seen from (6.30) that it is not possible to recover $L = A_j^{-1} \cdot l_j$ and the distance from the line to the origin $\|N_j\|$ – which would have characterized the Euclidean parameters of the line [6.30] – from the equations, without knowing the A-matrix. We, in fact, recover l_j only up to a scale factor.

However, we can write for a point m_i on the line d_i that there is a correspondent m'_j on the line d_j with $m'_j = m_j + \lambda\, l_j$ for some unknown λ. Replacing in (6.3), using (6.30) and eliminating Z_j and λ we derive, after some straightforward simplifications, the depth Z_i^a of a point $m_i^a \in d_i$, as

$$ Z_i^a \left(m_i^a{}^T \cdot Q_{ij} \cdot n_j \right) + \left(s_{ij}^T \cdot n_j \right) = 0 \ . $$

This equation, as expected, corresponds to the use of the retinal disparity in a direction normal to the line. Finally, we can compute the depth of any 3D point of the corresponding 3D line given its location on the 2D line, using only the Qs-representation. We recover the location of the line, up to a transformation, as discussed for points.

From Line Correspondences to Point Correspondences. We consider 2 points m^a and m^b on a 2D line d. In frame i the point m_i^a is in correspondence with a point in frame j of the form $m_j^a + \lambda^a\, l_j$, for some unknown λ^a, since we can not locate its correspondent on the 2D line, because we know it up to displacement on this 2D line. We thus must consider a "sliding" point. Similarly, a point m_i^b is in correspondence with a point of the form $m_j^b + \lambda^b\, l_j$, for some unknown λ^b.

Let us now assume that we know these two "partial" correspondences. This corresponds to taking only the *normal displacements* of these points into account. We thus can constrain Q_{ij} and s_{ij} since we can apply (6.3) rewritten in the form

$$ \begin{cases} Z_i^a\, Q_{ij}^{-1} \cdot m_i^a & = \ Z_j^a \left(m_j^a + \lambda^a\, l_j \right) - s_{ji} \ , \\ Z_i^b\, Q_{ij}^{-1} \cdot m_i^b & = \ Z_j^b \left(m_j^b + \lambda^b\, l_j \right) - s_{ji} \ . \end{cases} $$

But from (6.26) we have $n_i = \rho_i \left(m_i^a \wedge m_j^b \right)$ while $l_i = \kappa_i (Z_i^b\, m_i^b - Z_i^a\, m_j^a)$. Using these relations, we can combine with the previous equations and obtain after some manipulations

$$ \begin{cases} \left\{ \dfrac{Z_j^a\, Z_j^b \left[1 + \left(Z_i^b\, \lambda^b - Z_i^a\, \lambda^a \right) \kappa_j \right]}{\rho_j} \right\} n_j \\ \qquad = \left[\dfrac{Z_i^a\, Z_i^b\, \det(Q_{ji})}{\rho_i} \right] Q_{ji}^T \cdot n_i + \left(Z_i^b\, \lambda^b - Z_i^a\, \lambda^a - \dfrac{1}{\kappa_j} \right) s_{ji} \wedge l_j \ , \\ l_i = k\, Q_{ij}\, l_j \end{cases} $$

which, in turn, can be identified with (6.28) and (6.29), after some additional algebra.

In other words, if we know these two "partial" correspondences, we can predict the "uncalibrated" retinal motion of the line. One important consequence is that when *considering any line we can replace its contribution to the evaluation of the uncalibrated motion by two normal correspondences between two points of the line.* We can summarize this result as follows:

Proposition 6.3. The retinal motion of a line is entirely determined by the normal displacement of two points of this line.

This result is very important for implementations as already stress in [6.22] for other estimations, and will be used in the implementation proposed in this book.

6.3.2 Motion of Lines in Three Views

Let us now study the problem of computing the motion parameters S_{ij}, s_{ij} and q_{ijk}, considering points and/or lines correspondences.

In this section we are going to study the case of three views in detail, for several reasons: (1) It is known that for lines, when calibration is given, 3 views are needed [6.2, 4], while we now know that at least 3 views are also needed considering point correspondences when calibration is not given. (2) The geometric approach focus on the three views problem. (3) In the literature, most of the results are derived considering 2 or 3 views only, and this will allow to compare with previous works.

Reviewing Motion Equations for Points. To obtain these equations, we simply eliminate parameters related to the 3D structure of the scene.

When point correspondences are given, eliminating Z_0, Z_1 and Z_2 in (6.3) and using (6.19), we have

$$m_1^T \cdot \underbrace{\tilde{s}_{10} \cdot Q_{10}}_{F_{10}} \cdot m_0 \;=\; 0 \;;$$

$$m_1^T \cdot \underbrace{\tilde{s}_{12} \cdot Q_{12}}_{F_{12}} \cdot m_2 \;=\; 0 \;; \qquad\qquad (6.32)$$

$$m_0^T \cdot \underbrace{Q_{10}^T \cdot (\tilde{s}_{12} - \tilde{s}_{10}) \cdot Q_{12}}_{F_{02}} \cdot m_2 \;=\; 0 \;.$$

Please note that these three equations are independent, as already shown [6.25].

The two first equalities can be written, equivalently,[4]

$$0 = m_1 \wedge \left\{ \left[\underbrace{s_{10} \wedge (Q_{10} \cdot m_0)}_{F_{10} \cdot m_0} \right] \wedge \left[\underbrace{s_{12} \wedge (Q_{12} \cdot m_2)}_{F_{12} \cdot m_2} \right] \right\} \qquad (6.33)$$

and the last one

$$|(Q_{10} \cdot m_0), s_{12}, (Q_{12} \cdot m_2)| = |(Q_{10} \cdot m_0), s_{10}, (Q_{12} \cdot m_2)| \;. \qquad (6.34)$$

This means that given a point correspondence $(m_0 \leftrightarrow m_2)$ in two views, one can – knowing the Q-matrices and s-vectors – predict the location of the

---

[4] In fact $m_1 \wedge X = 0 \Leftrightarrow m_1 = \lambda X$ for some λ. Therefore, m_1 is entirely defined from the correspondence m_0 and m_2, using (6.33), since it is an homogeneous quantity, as obtained in [6.31].

point in the third view. This means that we can perform "trinocular stereo, when the system is weakly calibrated" [6.31, 32], that is without recovering explicitly the Euclidean parameters of the system.

These results, considering point correspondences were already known. The corresponding results considering lines are not, and we derive them now.

Motion Equations for Lines. Eliminating α_1, α_0, α_2 and $\beta_1/\det(A_1)$ in (6.28), written for $(j = 1, i = 0)$ and $(j = 1, i = 2)$ we derive

$$0 = n_1 \wedge \left[\frac{Q_{01}^T \cdot n_0}{(s_{01}^T \cdot n_0)} - \frac{Q_{21}^T \cdot n_2}{(s_{21}^T \cdot n_2)} \right] . \tag{6.35}$$

Then each line correspondence provides at least two equations, as in the calibrated case, while knowing a line correspondence between two views allows to compute the location of the line in the third view, because this equation allows to define n_1, when knowing n_0 and n_2, up to a scale factor.

This shows that correspondences can be verified without recovering the Euclidean parameters, as in the case of points [6.32].

This equation can easily be decomposed in the two equivalent scalar equations:

$$|n_1, Q_{01}^T \cdot n_0, Q_{21}^T \cdot n_2| = 0 \; ; \; \frac{\|n_1 \wedge Q_{01}^T \cdot n_0\|}{(s_{01}^T \cdot n_0)} = \frac{\|n_1 \wedge Q_{21}^T \cdot n_2\|}{(s_{21}^T \cdot n_2)} \tag{6.36}$$

which correspond to the two equations found by Liu and Huang [6.2] in the case where calibration is given. They are now generalized.

In the Euclidean case, the geometrical interpretation of the first equation is that the three normals of the three planes P_0, P_1 and P_2 (Fig. 6.6) are coplanar [6.30], and using the Qs-interpretation we can easily verify, that this interpretation is still valid. In the Euclidean case, the geometrical interpretation of the second equation is that the distance from the line to the origin, computed by triangulation using view 0 and 1 is the same as the distance from the line to the origin, computed by triangulation using view 2 and 1 [6.29], but this interpretation is no more valid because the related concept is only Euclidean.

We can permute the indices 0, 1 and 2 in (6.35) and obtain two other equations which relate n_0, n_1 and n_2, but contrary to the case of points, we will show that they are equivalent.

Introducing the decomposition of a Q-matrix from (6.12) in (6.35) yields

$$0 = n_1 \wedge \left(\frac{S_{01}^T \cdot n_0}{s_{01}^T \cdot n_0} - \frac{S_{21}^T \cdot n_2}{s_{21}^T \cdot n_2} + \underbrace{r_{01} - r_{21}}_{q_{012}} \right) . \tag{6.37}$$

This shows that we cannot recover all parameters of the Q-matrices, but only S-matrices, s-vectors and q-vectors and leads to the following result:

Proposition 6.4. Considering correspondences between lines the retinal motion is parameterized using the Ssq-representation, i.e. 18 parameters for 3 views, as for points and only two equations are available for each line correspondence.

Proof. Let us multiply Q_{ij} by λ_{ij} and s_{ij} by μ_{ij} in (6.35) such that

$$\frac{\lambda_{01}}{\mu_{01}} = \frac{\lambda_{21}}{\mu_{21}} .$$

The equations are not modified. Therefore $Q_{01}, Q_{21}, s_{01}, s_{21}$ are only recovered up to three scale factors, leaving three parameters unknown.

Moreover the equations on lines are only function of $q_{012} = r_{01} - r_{21}$ whereas the sum of these two vectors can not be found. Yet another triplet of parameters unknown. We thus do not have an explicit dependency with respect to 24 parameters, but at most $24 - 3 - 3 = 18$.

Let us now show that (6.35) is equivalent to any other equations relating n_0, n_1 and n_2. Since the only way to generate an equation about the motion of the line which is independent of the structure of the line is to eliminate the parameters related to this structure, i.e. α_i and β_i, the only way to generate an equation about the motion of line between three views from (6.28), is to generate an equation of the form of (6.35), up to a permutation of the indices.

Let us consider, for instance, the equation

$$0 = n_2 \wedge \left[\frac{Q_{12}^{\mathrm{T}} \cdot n_1}{(s_{12}^{\mathrm{T}} \cdot n_1)} - \frac{Q_{02}^{\mathrm{T}} \cdot n_0}{(s_{02}^{\mathrm{T}} \cdot n_0)} \right]$$

obtained after a permutation of the indices. Any other permutation could have been chosen.

Multiplying on the left by Q_{21} yields

$$\lambda\, n_1 = a\, Q_{21}^{\mathrm{T}} \cdot n_2 + b\, Q_{01}^{\mathrm{T}} \cdot n_0 \quad ; \quad \frac{\lambda}{b} = \frac{s_{02}^{\mathrm{T}} \cdot n_0}{s_{12}^{\mathrm{T}} \cdot n_1} .$$

But $(s_{02}^{\mathrm{T}} \cdot n_0) = (s_{12}^{\mathrm{T}} \cdot Q_{01}^{\mathrm{T}} \cdot n_0) - (s_{10}^{\mathrm{T}} \cdot Q_{01}^{\mathrm{T}} \cdot n_0)$ and $\lambda\, (s_{12}^{\mathrm{T}} \cdot n_1) = b\, (s_{12}^{\mathrm{T}} \cdot Q_{01}^{\mathrm{T}} \cdot n_0) + a\, (s_{12}^{\mathrm{T}} \cdot Q_{21}^{\mathrm{T}} \cdot n_2)$. We thus obtain

$$\frac{a}{b} = -\frac{s_{01}^{\mathrm{T}} \cdot n_0}{s_{21}^{\mathrm{T}} \cdot n_2}$$

and as a consequence (6.28). But this derivation is true for any permutation of the indices, thus can be done in the reverse way. These two equations are equivalent, and so are any equation about the motion of lines using three views.   □

Therefore, the same motion parameters can be recovered considering either point correspondences or line correspondences or both. More precisely, the following minimum numbers of tokens in generic positions needed for the estimation are[5]

---

[5] The fractional numbers mean that among the three equations provided by a each point correspondence, one or two is not mandatory.

Lines	0	1	2	3	4	5	6	7	8	9
Points	6	$6 - \frac{2}{3}$	$5 - \frac{1}{3}$	4	$4 - \frac{1}{3}$	$4 - \frac{2}{3}$	3	$2 - \frac{2}{3}$	$1 - \frac{1}{3}$	0

and we note that, surprisingly perhaps, a line correspondence leads to 2 equations only, whereas a point correspondence leads to 3 equations, all this in the general case.

We now would like to clarify why we have this identical representation for points and lines. A way to interpret the reason behind this algebraic identity is to try to construct a geometrical object which allows to understand these relationship. For instance, considering two views, this geometrical object is the epipolar geometry. It appears that another geometrical object exists in the present case which generalizes the epipolar geometry: this is the trifocal geometry as presented now.

6.3.3 Conclusion on Lines and Points Motion

We have now established that the same parameterization can be used to estimate the motion of lines and/or points, in the general case, where no calibration is given.

Note that each line correspondences bring two equations for the first triple of views and two equations for each additional view, i.e. $2(N-1)$ equations along the image sequence. Let us consider a sequence of $N+1$ views, Ln correspondences between lines and Pt correspondences between points, in order to solve the system, we must have

$$[2(N-1)] \, Ln + (2N-1) \, Pt \leq 11 \, N - 4 \,;$$

therefore at least 9 lines for 3 views, and 8 lines for 4 views, 7 lines for 5, 7 or 8 views, 6 lines for 9 views or more. We need always more than 5 line correspondences. If we combine points and lines we can summarize the minimal number of views required to analyse the projected motion in the following table in which configurations for which several solutions might exist have been quoted by a *. This corresponds to situations were points and/or lines provide just the minimal number of equations. It is clear from this table that: (1) as soon as a few points and lines are given, we can recover the projected motion parameters using three views; (2) using line correspondences allows to disambiguate the estimation of motion, and allows to decrease the number of required points.

Ln Pt	0	1	2	3	4	5	6	7	8	9	10
0							9*	5	4	3*	3
1						8*	4*	4	3	3	3
2					7*	4	3*	3	3	3	3
3				6*	4	3	3	3	3	3	3
4			5*	3*	3	3	3	3	3	3	3
5		4*	3	3	3	3	3	3	3	3	3
6	3*	3	3	3	3	3	3	3	3	3	3
7	2*	2*	2*	2*	2*	2*	2*	2*	2*	2*	2*

When establishing this result, we have also studied a class of linear algorithms to estimate this representation. We have made obvious the fact that it is very difficult to manage linear algorithms in this case, because the estimated parameters verify an important number of complicated constraints. Previous studies on the computation of motion from point or line correspondences, when no calibration, do not integrate all these constraints [6.5, 9] although the existence of these constraints has been already established and used for point correspondences [6.25].

The generalization to a sequence of views is now done and at the implementation level, we can avoid this level of complexity and build the algorithms on local correspondences only. This is discussed in the next section.

6.4 Implementation and Experimental Results

6.4.1 Implementation of the Motion Module

Let us now discuss the implementation of a module which takes point or token correspondences as input, and output an estimation of the affine motion parameters.

Thanks to several other developments in the field such as [6.3, 12, 21, 30, 33] we do not have to discuss again how to implement such a module in great details but simply can base our work on previous experiences. The main features to be taken into account are the following:

- Point or token correspondences are always defined with a certain uncertainty, often represented by a covariance matrix, and estimation criteria must weight their estimates using this uncertainty.
- It is always more robust and reliable to have a criterion based on a retinal measurement error (i.e. a retinal disparity or a image related quantity), even in the uncalibrated case [6.33], because this quantity corresponds to the physical measure. Obviously, when the retinal disparity has been canceled, all information about the motion has been extracted.

– It is always possible – and sometimes more efficient [6.30, 34] – to compute motion *and* structure at the same time, instead of eliminating structure parameters to estimate motion and then structure from motion. We will follow this track here.

Using Retinal Disparities as Measurement Error. Let us consider a set of P point correspondences in a sequence of $N+1$ views. We index the points with $k \in \{1 \dots P\}$ and the different views with $i \in \{0 \dots N\}$.

The main source of error, considering such point correspondences, is related to the uncertainty in the pixel localization and the uncertainty in the correspondence, the latter being more important than the former. A simple way to represent this uncertainty is to consider for each point m_i^k that the value of m_j^k is given with a certain uncertainty represented using the inverse of a covariance matrix, also called *information matrix*:

$$\left(\Lambda_i^k \right)^{-1} = \begin{pmatrix} I_{uu} & I_{uv} \\ I_{uv} & I_{vv} \end{pmatrix} .$$

In other words, we consider that for a given m_j^k, the corresponding measure is $m_i^k + \boldsymbol{\nu}_i^k$, while $\boldsymbol{\nu}_j^k$ is a Gaussian additive noise of covariance Λ_i^k.

Dealing with Non-Punctual Correspondences. In practice, we not always assume that this symmetric positive matrix is definite i.e. invertible. On the contrary, if the retinal disparity is defined in only one direction, say $\boldsymbol{n}_i^k = (\cos(\varsigma), \sin(\varsigma))^{\mathrm{T}}$, whereas in the orthogonal direction $\boldsymbol{t}_i^k = (-\sin(\varsigma), \cos(\varsigma))^{\mathrm{T}}$ the retinal correspondence is undefined, we can integrate this constraint very easily by defining an information matrix of the form

$$\left(\Lambda_i^k \right)^{-1} = \frac{1}{(\sigma_i^k)^2} \boldsymbol{n}_i^k \cdot \boldsymbol{n}_i^{k\,\mathrm{T}} . \tag{6.38}$$

Its interpretation is that the variance of the measure in the direction where the retinal correspondence is undefined is infinite (notion of generalized inverse) [6.16].

A 2D correspondence between a point m_i on a 2D curve and another point m_j on the corresponding 2D curve after a retinal displacement can be integrated in our framework. Along a curve (see Fig. 6.7), the tangential displacement is generally ambiguous (aperture problem) [6.16], except for points with a high curvature (corners, junctions). Let $\boldsymbol{t}_i^k$, $\|\boldsymbol{t}_i^k\| = 1$, be the tangent of the curve on m, and $\boldsymbol{n}_i^k$, $\|\boldsymbol{n}_i^k\| = 1$.

In this situation, the planar Euclidean displacement δm is obtained in the normal direction only, written $\beta = (\boldsymbol{n}_i^{k\,\mathrm{T}} \cdot \delta m)$, whereas $(\boldsymbol{t}_i^{k\,\mathrm{T}} \cdot \delta m)$ is undefined. The uncertainty for this equation is easily computable. Since only the disparity in the direction of $\boldsymbol{n}$ influences the measure, we can calculate the uncertainty as $(\sigma_i^k)^2 = \boldsymbol{n}_i^{k\,\mathrm{T}} \cdot \Lambda_i^k \cdot \boldsymbol{n}_i^k$, where Λ_i^k is the covariance of the correspondence. We can integrate this situation in our framework, by simply choosing a information matrix, as given in (6.38).

The case of line correspondences is a particular case for which only two correspondences are to be used, as discussed now.

From Line Correspondences to Point Correspondences. Let us consider a 2D line d_i with equation $m \in d_i \Leftrightarrow 0 = \boldsymbol{n}_i^{\mathrm{T}} \cdot m_i$ represented by an homogeneous vector $\boldsymbol{n}_i$ in correspondence with a 2D line d_j represented by an homogeneous vector $\boldsymbol{n}_j$. We reduce this equation on line correspondences to equations involving points correspondences, using Prop. 6.3, as developed now.

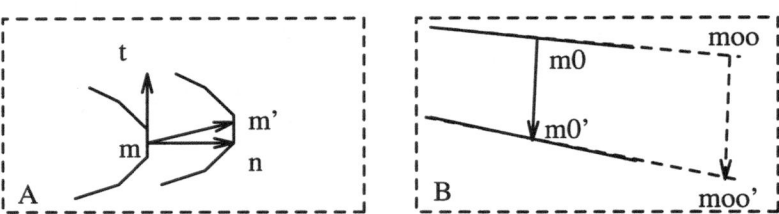

Fig. 6.7. Using correspondences between non punctual primitives; (**a**) correspondence related to the normal motion of a curve, (**b**) correspondence related to a line segment

Let us choose two points m_0 and m_1 on this 2D line. In practice, we are going to detect line-segments in an image. It has been shown [6.21] that we better choose the center of the segment m_0 and the direction of the segment, or point at infinity $m_1 = m_\infty$ (see Fig. 6.7). This will minimize the covariance of the measures, as demonstrated in [6.35]. However, any other pair of points could be used.

Each point is in correspondence with a point for which we cannot determine the exact location but only its location up to a displacement on the line, i.e. a "sliding point".

In other words, if we have a line correspondence, *constraining the retinal motion using the equation on the line parameter $\boldsymbol{n}$ or using the normal correspondences between two points on this line is equivalent.* We thus do not need to implement specifically the case of lines, but can limit our calculations to the case of points, with appropriate information matrices.

Dealing with Local Correspondences. What we have implicitly assumed from now on is that each point has been matched *all along the sequence*. This is by no means a realistic assumption and we must be able to integrate the fact that a point correspondence is valid only during a few frames.

This is very easy to do, using our information matrices again. Let us consider that the point correspondences have been established from view j_1 to view j_2 only, then we set $(\Lambda_j^k)^{-1} = 0$ for all $j < j_1$ and $j \geq j_2$ and the related equations, which are undefined, will not be taken into account because their weight is zero.

This allows to carry on our study, considering P points in the $N+1$ views but with suitable information matrices which integrate the fact that not all correspondences are established all along the sequence.

Parameterization of the Uncalibrated Motion and Structure. Let us consider again a sequence of $N+1$ views and P point-correspondences in this sequence.

We are going to represent the uncalibrated motion in the sequence by Q-matrices and s-vectors between *two consecutive frames*, i.e. $Q_{i+1\,i}$ and $s_{i+1\,i}$. All other Q-matrices and s-vectors are entirely defined by these quantities, using (6.4).

As discussed in this book, the Qs-representation is not entirely determined in the image sequence but is only determined up to:

(1) *A global scale factor for the image sequence.* We fix this scale factor by setting $\|s_{10}\| = \sigma_s$, i.e. by constraining the uncalibrated translation in the two first consecutive images. A reasonable default value is $\sigma_s = 1$. This corresponds to a reconstruction of the depths up to a scale factor as always the case in a monocular image sequence.

(2) *One r-vector in the sequence.* We fix $r_{10} = (\sigma_0, \sigma_1, \sigma_2)^{\mathrm{T}}$. A reasonable default value is $(\sigma_0, \sigma_1, \sigma_2) = (0, 0, 1)$. This corresponds to assuming that the plane at infinity is fronto-parallel, an affine retinal frame of reference being taken into account.

(3) *An expansion factor between each pair of consecutive frames.* We note the expansion factor between frame i and frame $i + 1$: $\{\mathcal{Z}_{i+1\,i}\}_{i=0...N-1}$. A reasonable default value is $\forall i, \mathcal{Z}_{i+1\,i} = 1$, especially when the camera is not performing a zoom, since it corresponds to assuming that the relative scales between two consecutive frames are left unchanged. This scale factor indetermination is avoided by simply choosing $Q_{i+1\,i}^{22} = 1$, i.e. fix the last component of each Q-matrix.[6]

We can either determine these parameters from other perceptual processes (Euclidean calibration, odometric cues, etc.) or set them to the proposed default values. In that case, we are not going to estimate the "true" parameters but the depths and displacements up to some affine and projective transformations as discussed in this book. In this implementation, we take these parameters as inputs of our algorithm.

Parameterization in the first view: we use a vector $\mu_0 \in \mathcal{R}^7$.

For $n = 0$: $\mu_0 = (\theta_0, \phi_0, a_0, b_0, d_0, e_0, f_0)^{\mathrm{T}} \in \mathcal{R}^7$

such that

---

[6] This constraint is not far from the reality since in an image sequence we expect the rotational component of the motion between two frames to be rather small and the calibration matrix not to be much modified. As a consequence, the matrix $Q_{i+1\,i} = A_{i+1} \cdot R_{i+1\,i} \cdot A_i^{-1}$ is expected to be close to the identity.

$$\begin{cases} \boldsymbol{s}_{10} &= \sigma_s \left(\cos(\theta_0)\cos(\phi_0), \sin(\theta_0)\cos(\phi_0), \sin(\phi_0)\right)^{\mathrm{T}}, \\ Q_{10} &= S_{10} + \boldsymbol{s}_{10} \cdot \begin{pmatrix} \sigma_0 \\ \sigma_1 \\ \sigma_2 \end{pmatrix}^{\mathrm{T}} \\ \text{with} & \\ S_{10} &= \begin{pmatrix} \cos(\theta_0)\sin(\phi_0) \\ \sin(\theta_0)\sin(\phi_0) \\ -\cos(\phi_0) \end{pmatrix} \cdot \begin{pmatrix} a_0 \\ b_0 \\ c_0 \end{pmatrix}^{\mathrm{T}} \\ & + \begin{pmatrix} -\sin(\theta_0) \\ \cos(\theta_0) \\ 0 \end{pmatrix} \cdot \begin{pmatrix} d_0 \\ e_0 \\ f_0 \end{pmatrix}^{\mathrm{T}}, \\ c_0 &= \dfrac{\sigma_2 \sigma_s \sin(\phi_0) - 1}{\cos(\phi_0)}. \end{cases}$$

This parameterization does not cover all configurations and is undefined if $s_{10}^0 = s_{10}^1 = 0$. But, if we use two other maps obtained by permuting the trigonometric factors in the definition of $\boldsymbol{s}_{10}$, we parameterize all configurations. This s well known, each we consider unary vectors, which cover the unary sphere. With this technic we can construct a chart of our representation using 2 maps.

This mechanism also provides a parameterization of the F-matrix, and we have from (6.11)

$$F_{10} = \begin{pmatrix} -\sin(\theta_0) \\ \cos(\theta_0) \\ 0 \end{pmatrix} \cdot \begin{pmatrix} a_0 \\ b_0 \\ c_0 \end{pmatrix}^{\mathrm{T}} - \begin{pmatrix} \cos(\theta_0)\sin(\phi_0) \\ \sin(\theta_0)\sin(\phi_0) \\ -\cos(\phi_0) \end{pmatrix} \cdot \begin{pmatrix} d_0 \\ e_0 \\ f_0 \end{pmatrix}^{\mathrm{T}}.$$

As expected, $Q_{10}^{22} = 1$, $\boldsymbol{s}_{10}^{\mathrm{T}} \cdot S_{10} = 0$, $\|\boldsymbol{s}_{10}\| = \sigma_s$ and $\left(Q_{10}^{\mathrm{T}} \cdot \boldsymbol{s}_{10}\right)/\|\boldsymbol{s}_{10}\|^2 = \boldsymbol{r}_{10} = (\sigma_0, \sigma_1, \sigma_2)^{\mathrm{T}}$. We thus verify the required constraints. Please note that c_0 as been chosen so that $Q_{10}^{22} = 1$.

Parameterization in the other views: we use vectors $\boldsymbol{\mu}_n \in \mathcal{R}^{11}$.

For $\quad 0 < n < N$:

$$\boldsymbol{\mu}_n = \left(s_{i+1\,i}^0, s_{i+1\,i}^1, s_{i+1\,i}^2,\right.$$
$$\left. Q_{i+1\,i}^{00}, Q_{i+1\,i}^{01}, Q_{i+1\,i}^{02}, Q_{i+1\,i}^{10}, Q_{i+1\,i}^{11}, Q_{i+1\,i}^{12}, Q_{i+1\,i}^{20}, Q_{i+1\,i}^{21}\right)^{\mathrm{T}} \in \mathcal{R}^{11}$$

while again, we choose $Q_{i+1\,i}^{22} = 1$.

By doing this we use $7 + 11\,(N-1) = 11\,N - 4$ parameters, as expected, and verify all the constraints of this representation.

Parameterization of the 3D depths: the structure of the scene is very simply parameterized by the inverse of the affine depths of the points, called *proximity*:

$$\pi_i^k = \frac{1}{Z_i^k}$$

for a point $k \in \{1\ldots p\}$ in frame $i \in \{0\ldots N\}$.

A Criterion to Estimate the Qs-Representation. Let us now define the measurement equations. As stated in this book, all measurement equations, considering point correspondences, are given by (6.5). In this equation, we have decomposed the equations in (a) a prediction of the retinal disparity and (b) a prediction of the depth. This is exactly the decomposition we need, since we would like:

(1) to minimize the retinal disparities, between consecutive frames:

$$\epsilon_i^k(\mu_i, \pi_i^k) = \begin{vmatrix} u_{i+1} & - & \dfrac{Q_{i+1\,i}^{00}\, u_i + Q_{i+1\,i}^{01}\, v_i + Q_{i+1\,i}^{02} + \frac{1}{Z_i^k}\, s_{i+1\,i}^0}{Q_{i+1\,i}^{20}\, u_i + Q_{i+1\,i}^{21}\, v_i + Q_{i+1\,i}^{22} + \frac{1}{Z_i^k}\, s_{i+1\,i}^2} \\[2mm] v_{i+1} & - & \dfrac{Q_{i+1\,i}^{10}\, u_i + Q_{i+1\,i}^{11}\, v_i + Q_{i+1\,i}^{12} + \frac{1}{Z_i^k}\, s_{i+1\,i}^1}{Q_{i+1\,i}^{20}\, u_i + Q_{i+1\,i}^{21}\, v_i + Q_{i+1\,i}^{22} + \frac{1}{Z_i^k}\, s_{i+1\,i}^2} \end{vmatrix}$$

weighted by the inverse of their covariances $(\Lambda_i^k)^{-1}$,

(2) but also the errors between predicted and estimated depths, i.e.

$$\xi_i^k(\mu_i, \pi_{i+1}^k, \pi_i^k) = \frac{1}{Z_{i+1}^k} - \frac{1}{Z_i^k}\, \frac{Z_{i+1\,i}}{Q_{i+1\,i}^{20}\, u_i + Q_{i+1\,i}^{21}\, v_i + Q_{i+1\,i}^{22} + \frac{1}{Z_i^k}\, s_{i+1\,i}^2}$$

in which we have introduced the zoom factor $Z_{i+1\,i}$. The quantity $\xi_i^k(\mu_i, \pi_{i+1}^k, \pi_i^k)$ is weighted by its information $V_{\xi_i^k}^{-1}$ computed as follows:[7]

$$V_{\xi_i^k}^{-1} = \frac{\partial \pi_i^k}{\partial \xi_i^k}^{\mathrm{T}} \cdot V_{\pi_i^k}^{-1} \cdot \frac{\partial \pi_i^k}{\partial \xi_i^k},$$

$$V_{\pi_i^k}^{-1} = \frac{\partial \epsilon_i^k}{\partial \pi_i^k}^{\mathrm{T}} \cdot (\Lambda_i^k)^{-1} \cdot \frac{\partial \epsilon_i^k}{\partial \pi_i^k},$$

$$\frac{\partial \pi_i^k}{\partial \xi_i^k} = \frac{Q_{i+1\,i}^{20}\, u_i + Q_{i+1\,i}^{21}\, v_i + Q_{i+1\,i}^{22} + \frac{1}{Z_i^k}\, s_{i+1\,i}^2}{Z_{i+1\,i}},$$

$$\frac{\partial \epsilon_i^k}{\partial \pi_i^k} \simeq \frac{1}{Q_{i+1\,i}^{20}\, u_i + Q_{i+1\,i}^{21}\, v_i + Q_{i+1\,i}^{22} + \frac{1}{Z_i^k}\, s_{i+1\,i}^2}$$

$$\cdot \underbrace{\begin{pmatrix} s_{i+1\,i}^0 - u_{i+1}\, s_{i+1\,i}^2 \\ s_{i+1\,i}^1 - v_{i+1}\, s_{i+1\,i}^2 \end{pmatrix}}_{\boldsymbol{\nu}_i^k}$$

so that, finally:

$$V_{\xi_i^k}^{-1} = \frac{1}{Z_{i+1\,i}^2}\, \boldsymbol{\nu}_i^{k\,\mathrm{T}} \cdot (\Lambda_i^k)^{-1} \cdot \boldsymbol{\nu}_i^k.$$

---

[7] In an equation of the form $y = f(x)$ we can compute the inverse of the covariance of x from the inverse of the covariance of y from the well-known formula obtained applying the Implicit Function Theorem:
$$\Lambda_x^{-1} = \frac{\partial f}{\partial x}^{\mathrm{T}} \cdot \Lambda_y^{-1} \cdot \frac{\partial f}{\partial x}.$$

Combining these measures we obtain the following criterion:

$$J(\{\mu_i\}_{i=0\ldots N-1}, \{\pi_i^k\}_{k=1\ldots P, i=0\ldots N})$$

$$= \sum_{i=0}^{N-1}\sum_{k=1}^{P} \epsilon_i^{k\,\mathrm{T}} \cdot (\Lambda_i^k)^{-1} \cdot \epsilon_i^k + \sum_{i=0}^{N-1}\sum_{k=1}^{P} \xi_i^k \, V_{\xi_i^k}^{-1} \, \xi_i^k \, .$$

The problem of computing motion and structure in a monocular image sequence when no calibration is given has been reduced to the present optimization problem.

Implementation of the Minimization Process. In order to proceed to the minimization of this rather huge criterion, we decompose the minimization into two subproblems:

Motion from structure: We minimize with respect to the motion parameters $\{\mu_i\}_{i=0\ldots N-1}$, the structure parameters $\{\pi_i^k\}_{k=1\ldots P, i=0\ldots N}$ being constant. In that case the previous criterion is the sum of N independent terms:

$$J(\{\mu_i\}_{i=0\ldots N-1}, \{\pi_i^k\}_{k=1\ldots P, i=0\ldots N}) = \sum_{i=0}^{N-1} J_{\mu_i}(\mu_i, \{\pi_i^k\}_{k=1\ldots P, i=0\ldots N}) \, ,$$

$$J_{\mu_i}(\mu_i, \{\pi_i^k\}_{k=1\ldots P, i=0\ldots N}) = \sum_{k=1}^{P} \epsilon_i^{k\,\mathrm{T}} \cdot (\Lambda_i^k)^{-1} \cdot \epsilon_i^k + \xi_i^k \, V_{\xi_i^k}^{-1} \, \xi_i^k$$

and – considering $\{\pi_i^k\}_{k=1\ldots P, i=0\ldots N}$ is constant –, the problem reduces to the independent minimization of the N sub-criteria J_{μ_i}, each of them requiring the minimization of the sum of 3 P terms[8] with respect to the parameters of μ_i, either 7 (for $i = 0$) or 11 (for $i > 0$). Such a minimization is quite easy to perform.

Structure from motion: We minimize with respect to the structure parameters $\{\pi_i^k\}_{k=1\ldots P, i=0\ldots N}$, the motion parameters $\{\mu_i\}_{i=0\ldots N-1}$ being constant. In that case the previous criterion is the sum of P independent terms:

$$J(\{\mu_i\}_{i=0\ldots N-1}, \{\pi_i^k\}_{k=1\ldots P, i=0\ldots N})$$

$$= \sum_{k=1}^{P} J_{\pi_k}(\{\mu_i\}_{i=0\ldots N-1}, \{\pi_i^k\}_{i=0\ldots N}) \, ,$$

$$J_{\pi_k}(\{\mu_i\}_{i=0\ldots N-1}, \{\pi_i^k\}_{i=0\ldots N}) = \sum_{i=0}^{N-1} \epsilon_i^{k\,\mathrm{T}} \cdot (\Lambda_i^k)^{-1} \cdot \epsilon_i^k + \xi_i^k \, V_{\xi_i^k}^{-1} \, \xi_i^k$$

and – considering $\{\mu_i\}_{i=0\ldots N-1}$ is constant –, the problem reduces to the independent minimization of the P sub-criteria J_{π_k}, each of them requiring

---

[8] Each correspondence provides 2 equations (eventually 1 vanishes, if $(\Lambda_i^k)^{-1}$ is of rank 1) for the disparity and 1 for the depth prediction.

the minimization of the sum of $3 (N - 1)$ terms with respect to the $N + 1$ parameters $\{\pi_i^k\}_{i=0...N}$. Such a minimization is again very easy to perform.

Obviously, this decomposition dramatically simplifies the amount of calculations. In order to accelerate the convergence of the estimation process, we use a very common strategy [6.30] in which we decompose the estimation process into two *phases*, a (1) boot-strapping phase and (2) several steady-state phases:

During the **boot-strapping phase**, we attempt to obtain a initial estimate of a subset of the parameters in order to start the minimization as close as possible to the final estimate. We must obtain an estimation of motion, without any knowledge on the structure, which corresponds to the *estimation of the fundamental matrix* $F_{10}(\mu_0)$. The corresponding criterion can be derived exactly as the previous one, and we have, very briefly, from [6.33]:

$$J'(\mu_0) = \sum_{k=1}^{P} \xi_0^k V_{\xi_0^k}^{-1} \xi_0^k + \sum_{k=1}^{P} \xi_1^k V_{\xi_1^k}^{-1} \xi_1^k \,,$$

$$\xi_0^k = \frac{m_1^{\mathrm{T}} \cdot F_{10}(\mu_0) \cdot m_0}{\sqrt{[F_{10}(\mu_0) \cdot m_0]^{\mathrm{T}} \cdot U \cdot [F_{10}(\mu_0) \cdot m_0]}} \,,$$

$$V_{\xi_0^k}^{-1} = \frac{1}{\frac{\partial \xi_0^k}{\partial m_1} \cdot (\Lambda_i^k) \cdot \frac{\partial \xi_0^k}{\partial m_1}^{\mathrm{T}}} = \frac{[F_{10}(\mu_0) \cdot m_0]^{\mathrm{T}} \cdot U \cdot [F_{10}(\mu_0) \cdot m_0]}{[F_{10}(\mu_0) \cdot m_0]^{\mathrm{T}} \cdot (\Lambda_i^k) \cdot [F_{10}(\mu_0) \cdot m_0]} \,,$$

$$\xi_1^k = \frac{m_1^{\mathrm{T}} \cdot F_{10}(\mu_0) \cdot m_0}{\sqrt{[F_{10}(\mu_0)^{\mathrm{T}} \cdot m_1]^{\mathrm{T}} \cdot U \cdot [F_{10}(\mu_0)^{\mathrm{T}} \cdot m_1]}} \,,$$

$$V_{\xi_1^k}^{-1} = \frac{1}{\frac{\partial \xi_1^k}{\partial m_0} \cdot (\Lambda_i^k) \cdot \frac{\partial \xi_1^k}{\partial m_0}^{\mathrm{T}}}$$

$$= \frac{[F_{10}(\mu_0)^{\mathrm{T}} \cdot m_1]^{\mathrm{T}} \cdot U \cdot [F_{10}(\mu_0)^{\mathrm{T}} \cdot m_1]}{[F_{10}(\mu_0)^{\mathrm{T}} \cdot m_1]^{\mathrm{T}} \cdot (\Lambda_i^k) \cdot [F_{10}(\mu_0)^{\mathrm{T}} \cdot m_1]} \,,$$

$$\text{with} \qquad U = \begin{pmatrix} 1 & 0 & 0 \\ 0 & 1 & 0 \\ 0 & 0 & 0 \end{pmatrix}$$

following the method developed and discussed by previous scientists [6.25, 33]. These quantities can be interpreted as follows: (a) We have chosen as measurement errors ξ_0^k and ξ_1^k which are the Euclidean distances from a point to its expected epipolar line, (b) in a symmetric way, since we have average the previous quantity in both views, (c) and weighted these quantities by the uncertainty of each correspondence. Please refer to [6.33] for a discussion.

In any case, note that this estimation cannot be done using line correspondences (we use 2 views) but point correspondences only. This appears in the fact that we use the covariance matrix (Λ_i^k) and not its inverse which is undefined for normal correspondences. This is a limitation of the present implementation that we do not use line correspondences for the boot-strapping

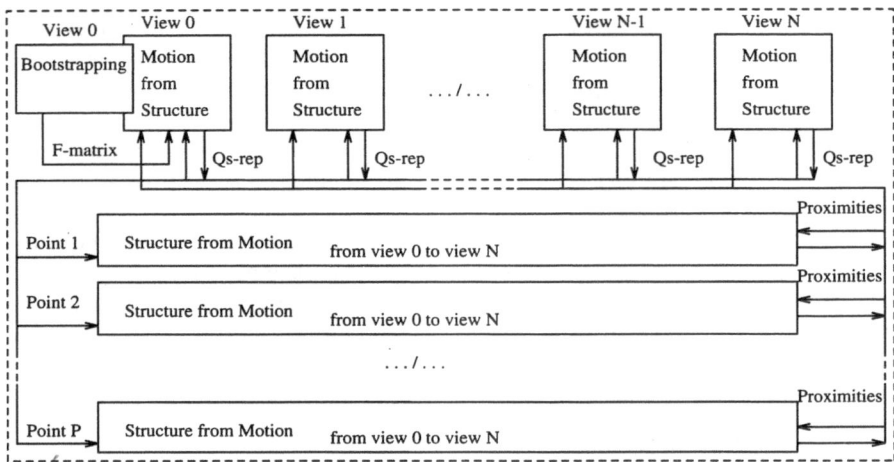

Fig. 6.8. Architecture of the minimization process

phase, although a straightforward implementation of the previous equations and a mechanism using three views is indeed possible.

The estimation of the fundamental matrix $F_{10}(\mu_0)$ provides an initial estimate of the motion parameters between view 0 and 1. From this initial estimate we can obtain a first value for the structure parameters from the "structure from motion" paradigm, minimizing the criteria $J_{\pi_k}(\{\mu_i\}_{i=0...N-1},$ $\{\pi_i^k\}_{i=0...N})$ with respect to $\{\pi_i^k\}_{i=0...N}$ for all k.

During the **steady-state phases**, we re-estimate alternatively "motion from structure" and "structure from motion", first considering frame 0 only, then integrating frame 0, 1, etc. When this mechanism has been executed up to view N, the estimation is finished.

This finally leads to a non-trivial architecture, illustrated in Fig. 6.8. We must report that we have tried and experimented several alternatives and variations of the proposed paradigm in order to obtain an efficient algorithm, for which we are going to report some experimental results in the next paragraph. This version of the implementation seems to be a good compromise between several constraints, not all being easy to formalize.

Fast Implementation of the Algorithm. In the proposed implementation, each minimization is performed using a comprehensive algorithm for finding the minimum of a sum of squares [6.36]. No derivatives are required, which is an advantage (less computations, adaptive estimation of either the Gauss–Newton directions or the Newton directions depending on the numerical stability of the algorithm, etc.) with respect to methods requiring the analytic gradient. The method is designed to ensure that steady progress is made whatever the starting point, and to have a rapid ultimate convergence, i.e. super-linear.

However this mechanism is not adapted to a real-time implementation for instance, but we can modified a bit our criterion to obtain a linear algorithm. Let us explain how.

If we consider the quantity:

$$\epsilon_i^k(\mu_i, \pi_i^k) = \left(Q_{i+1\,i}^{20} u_i + Q_{i+1\,i}^{21} v_i + Q_{i+1\,i}^{22} + \frac{1}{Z_i^k} s_{i+1\,i}^2 \right) \epsilon_i^k(\mu_i, \pi_i^k)$$

$$= \begin{vmatrix} \left(Q_{i+1\,i}^{20} u_i + Q_{i+1\,i}^{21} v_i + Q_{i+1\,i}^{22} + \frac{1}{Z_i^k} s_{i+1\,i}^2 \right) u_{i+1} \\ -Q_{i+1\,i}^{00} u_i + Q_{i+1\,i}^{01} v_i + Q_{i+1\,i}^{02} + \frac{1}{Z_i^k} s_{i+1\,i}^0 \\ \left(Q_{i+1\,i}^{20} u_i + Q_{i+1\,i}^{21} v_i + Q_{i+1\,i}^{22} + \frac{1}{Z_i^k} s_{i+1\,i}^2 \right) v_{i+1} \\ -Q_{i+1\,i}^{10} u_i + Q_{i+1\,i}^{11} v_i + Q_{i+1\,i}^{12} + \frac{1}{Z_i^k} s_{i+1\,i}^1 \end{vmatrix}$$

we can observe that this error is linear with respect to Q, s and the proximity $\pi = 1/Z$. Similarly, if we consider the quantity

$$\bar{\nu}_i^k = \left(\bar{Q}_{i+1\,i}^{20} u_i + \bar{Q}_{i+1\,i}^{21} v_i + \bar{Q}_{i+1\,i}^{22} + \frac{\bar{1}}{Z_i^k} \bar{s}_{i+1\,i}^2 \right)$$

estimated *a-priori*, i.e. the value of $\bar{\nu}_i^k$ is not determined by the algorithm, but input and constant, using a-priori information on motion and structure, the quantity

$$\zeta(\pi_{i+1}^k, \pi_i^k) = \left(Q_{i+1\,i}^{20} u_i + Q_{i+1\,i}^{21} v_i + Q_{i+1\,i}^{22} + \frac{1}{Z_i^k} s_{i+1\,i}^2 \right)$$

$$\cdot \xi_i^k(\mu_i, \pi_{i+1}^k, \pi_i^k)$$

$$= \bar{\nu}_i^k \frac{1}{Z_{i+1}^k} - \frac{1}{Z_i^k} Z_{i+1\,i}$$

is no more dependent on the motion, but only on the structure and is linear with respect to proximity.

As a consequence the modified criterion

$$J'(\{\mu_i\}_{i=0...N-1}, \{\pi_i^k\}_{k=1...P, i=0...N})$$

$$= \sum_{i=0}^{N-1} \sum_{k=1}^{P} \epsilon_i^{k\,T} \cdot \frac{(\Lambda_i^k)^{-1}}{(\bar{\nu}_i^k)^2} \cdot \epsilon_i^k + \sum_{i=0}^{N-1} \sum_{k=1}^{P} \zeta_i^k \frac{V_{\xi_i^k}^{-1}}{(\bar{\nu}_i^k)^2} \zeta_i^k$$

$$= \sum_{i=0}^{N-1} \underbrace{\sum_{k=1}^{P} \epsilon_i^{k\,T} \cdot \frac{(\Lambda_i^k)^{-1}}{(\bar{\nu}_i^k)^2} \cdot \epsilon_i^k}_{J'_{\mu_i}(\mu_i, \{\pi_i^k\}_{k=1...P, i=0...N})} + \sum_{i=0}^{N-1} \sum_{k=1}^{P} \zeta_i^k \frac{V_{\xi_i^k}^{-1}}{(\bar{\nu}_i^k)^2} \zeta_i^k$$

$$= \sum_{k=1}^{P} \underbrace{\sum_{i=0}^{N-1} \epsilon_i^{k\,T} \cdot \frac{(\Lambda_i^k)^{-1}}{(\bar{\nu}_i^k)^2} \cdot \epsilon_i^k + \zeta_i^k \frac{V_{\xi_i^k}^{-1}}{(\bar{\nu}_i^k)^2} \zeta_i^k}_{J'_{\pi_k}(\{\pi_i^k\}_{i=0...N})}$$

is, on the one hand, quadratic with respect to the components of the Q-matrix and s-vector and, on the other hand, quadratic with respect to proximity. Therefore, the minimization of $J'_{\mu_i}(\mu_i, \{\pi_i^k\}_{k=1...P, i=0...N})$ and $J'_{\pi_k}(\{\pi_i^k\}_{i=0...N})$ leads to linear normal equations.

With this strategy, each step of the minimization corresponds to solving a linear system of equations, except for μ_0 because of the trigonometric functions. Nevertheless, in this last case, the equations are linear with respect to $(a_0, b_0, d_0, e_0, f_0)$. As a consequence we can avoid using nonlinear minimization.

In the reality, we will verify, that considering the experimental results, the quantity $\bar{\nu}_i^k$ is not far from 1 so that the proposed approximation should be valid. It is so true, that even the original criterion yields "almost linear" normal equations. Therefore, the nonlinear minimization algorithm should converge in one or two steps. This will be verified in the next section.

We are not sure, depending on the application, that using linear equations is always a good choice because, although each step is rather quick to execute, the local minimization is less efficient and the global convergence might be affected, if we are far from the minimum. This problem is thus data dependent and will not be further discussed here.

6.4.2 Experimental Results

The goal of the present experimentation is just to show that, although the representation of motion in the uncalibrated case is somehow more complicated than what is usually done when calibration is available, it is possible to implement such a module as described before and obtained reasonable performances on real-data "from the lab".

Using a Method to Control the Obtained Values. We have chosen a monocular sequence of 9 frames containing 128 point correspondences. We had to choose an image sequence for which the results of the computation of the motion parameters (rigid motion + calibration) and structure parameters are computable using another well established method. Therefore we have chosen a sequence of views containing a calibration grid, for which all these parameters can be easily computed as often used in the past [6.25, 31].

A typical image sequence is shown in Fig. 6.9, for three consecutive images. The order of magnitude of the disparity between two frames is 2 to 4 pixels, corresponding to rotations of the object of about 5 deg. This corresponds to an acquisition at video rate.

For each frame, we have computed the usual calibration parameters, i.e the 3×3 matrix P_i and 3×1 vector $\boldsymbol{p}_i$ which defines the projection of a 3D point $M = (X, Y, Z)^{\mathrm{T}}$, given in an absolute frame of reference onto a retinal point $m_i = (u_i, v_i, 1)^{\mathrm{T}}$ using the relation [6.16]:

Fig. 6.9. Three images of the sequence used for the experimentation

$$Z_i \, m_i = P_i \cdot M + \boldsymbol{p}_i = [P_i \,|\, \boldsymbol{p}_i] \cdot \begin{pmatrix} X \\ Y \\ Z \\ 1 \end{pmatrix} .$$

If we write R_i and $\boldsymbol{t}_i$ for the rotation matrix and for the translation vector from the absolute frame of reference to the retinal frame of reference, we immediately verify that

$$P_i = A_i \cdot R_i \qquad ; \qquad \boldsymbol{p}_i = A_i \cdot \boldsymbol{t}_i$$

and the P-matrix must verify the following constraint $(P_i \cdot P_i^{\mathrm{T}})^{22} = 1$, since $P_i \cdot P_i^{\mathrm{T}} = A_i \cdot A_i^{\mathrm{T}} = K_i$ while $K_{ii}^{22} = 1$ from (6.7).

Now if we eliminate M in these equations taken in frame i and j we obtain

$$Q_{ij} = P_i \cdot P_j^{-1} \qquad ; \qquad \boldsymbol{s}_{ij} = \boldsymbol{p}_i - Q_{ij} \cdot \boldsymbol{p}_j .$$

In fact P_i can be considered as the collineation between the plane at infinity and the retinal plane corresponding to frame i, as easily verified considering a point $M = \|M\| (\boldsymbol{u}/\|\boldsymbol{u}\|)$ for huge $\|M\|$. The evolution of these parameters is also governed by our Qs-representation, since

$$\begin{cases} P_i &= Q_{ij} \cdot P_j \\ \boldsymbol{p}_i &= Q_{ij} \cdot \boldsymbol{p}_j + \boldsymbol{s}_{ij} \end{cases}$$

as obtained from the previous relation.

As a consequence, we have an estimate of all the motion and structure parameters using a well established method and can compare these results with the outputs of our module.

Experimental Measure of the External Parameters. As discussed in the previous section, we can estimate all parameters in this situation, and in particular estimate the external parameters which are inputs of our algorithms. We have obtained

$$\begin{aligned} \|\boldsymbol{s}_{10}\| &= && 17\,406.948\,594 \\ \boldsymbol{r}_{10} &= && (\text{ -3.3\,e-06 -5.4\,e-05 -0.001 }) \\ \{\mathcal{Z}_{i+1\,i}\}_{i=0...7} &= && \{\ 0.999\ 0.989\ 1.035\ 0.992\ 0.994\ 0.997\ 0.992\ 0.970\ 0.986\ \} \end{aligned}$$

which yields the following comments: (1) as expected, the zoom factors are very close to 1 (there were no zoom during this experimentation) and (2) the direction of the r-vector is very close to $(0, 0, 1)^{\mathrm{T}}$ also, since the plane at infinity is – in standard conditions – almost a fronto parallel plane. The magnitudes of s_{10} and r_{10} are of course very far to one, but that was also expected.

These magnitudes are well explained if we observe that a typical value of the Qs-representation is

$$
Q\_\{08\} = \begin{pmatrix} 0.959767 & 0.009758 & -27.106785 \\ -0.005585 & 0.945462 & 19.148278 \\ 0.000068 & 0.000002 & 0.981526 \end{pmatrix} \quad s\_\{08\} = \begin{pmatrix} 23802.317544 \\ -20696.781141 \\ -58.490636 \end{pmatrix}
$$

and it is obvious that the normalization based on the constraint $Q_{i+1\,i}^{22} = 1$ does not modify the order of magnitude of the obtained quantities.

However it appears that the obtained quantities have not the same order of magnitude and this can lead to some numerical instabilities in the algorithms. We have avoided such a problem by taking this order of magnitude into account in the numerical routines so that quantities expected to be very small or very high have been roughly normalized.

Let us now verify if our assumption about the fact that, considering a fast implementation, we can consider

$$
\bar{\nu}_i^k = \left(Q_{i+1\,i}^{20}\, u_i + Q_{i+1\,i}^{21}\, v_i + Q_{i+1\,i}^{22} + \frac{1}{Z_i^k}\, s_{i+1\,i}^2 \right)
$$

as almost constant is valid. Here are the values obtained along this typical image sequence:

$$
\begin{aligned}
\bar{\nu}_0^k &= +0.000021\, u^k - 0.000021\, v^k + 0.999955 - 5.6\,\mathrm{e} + 01\, \pi^k \\
\bar{\nu}_1^k &= +0.000045\, u^k - 0.000003\, v^k + 0.989326 - 9.3\,\mathrm{e} + 00\, \pi^k \\
\bar{\nu}_2^k &= -0.000189\, u^k + 0.000017\, v^k + 1.035403 + 5.0\,\mathrm{e} + 01\, \pi^k \\
\bar{\nu}_3^k &= +0.000022\, u^k + 0.000008\, v^k + 0.992263 - 4.2\,\mathrm{e} + 00\, \pi^k \\
\bar{\nu}_4^k &= +0.000026\, u^k - 0.000004\, v^k + 0.994453 - 1.5\,\mathrm{e} + 00\, \pi^k \\
\bar{\nu}_5^k &= +0.000017\, u^k - 0.000007\, v^k + 0.997433 - 9.0\,\mathrm{e} + 00\, \pi^k \\
\bar{\nu}_6^k &= +0.000033\, u^k - 0.000003\, v^k + 0.992366 - 1.8\,\mathrm{e} + 01\, \pi^k \\
\bar{\nu}_7^k &= +0.000092\, u^k + 0.000020\, v^k + 0.970632 - 1.2\,\mathrm{e} + 00\, \pi^k
\end{aligned}
$$

for values of π^k of about 10^{-3}, while $(u^k, v^k) \in [0 \ldots 512[\times [0 \ldots 512[$. It is clear that this quantity, although not constant, has a value very close to one so that the modified criterion proposed in as a fast implementation of our original criterion is realistic.

We thus have verified that the different assumptions made on the order of magnitude or the approximate values correspond to what is effectively measured in a real monocular sequence.

Performances of the Algorithm. Let us also very briefly report how *fast* (or slow!) our implementation was.

The "motion from structure" modules converge in about 16 iterations for μ_0, and only 2 to 3 iterations for the other motion parameters $\mu_n, n > 0$ and the convergence is obtain, from scratch, in about 1.5 s CPU-time on a Sun-II for 9 views and 128 points.

The "structure from motion" modules converges in about 5 to 10 iterations and the convergence is obtain, from scratch, in about 4.5 s CPU-time on a Sun-II for 9 views and 128 points.

When activated together these two sets of modules take about 3 s to 5 s CPU-time on a Sun-II for 9 views and 128 points (quicker because we never compute from scratch except the first module) for each iteration and the global convergence is obtained after 5 to 10 calls to each modules. In fact if we iterate a huge amount of time, we still gain something but this is peanuts. If we start close to the expected value (as in a steady-state mode) the convergence is obtain after 2 to 3 calls.

As a consequence, it is clear that the fact we have used an almost quadratic criterion for all parameters except the μ_0 yields to a very fast convergence, even if to maximize the robustness of the estimate we have not used the fast implementation.

We have not yet optimized the code thus CPU-times are still quite long, and must not be taken as definitive values. Previous experience in the field let us predict that a careful coding can decrease these times by a factor of 2 to 10.

Precision of the Algorithm, Simulation. Since we had an image sequence of real data, considering a set of corners on a calibration grid, we have also simulated the projection of these points onto the camera, using the numerical values obtained from the real image sequence as data for the simulation. As a consequence, we expect to have "realistic values" for the simulation.

We have added a white Gaussian noise to the retinal locations of each points, the standard deviation of this noise being $\sigma_\mathcal{N}$. We have considered that the precision on the retinal location of the points could have an uncertainty from 0.1 pixel (corresponding to accurate subpixel corner estimators for instance [6.37, 38]) up to 1 or 2 pixels for standard feature detectors. We have verified the stability of the algorithm also considering large errors of 5 pixels because this corresponds to the error amplitudes when false matchings occur in an image sequence [6.11, 12].

The precision of the estimate is measured considering the distance between the 11 components of the expected and estimated Qs-representations in the parameter space.

We have obtained the following experimental results:

Noise Level	Motion from Structure	Structure from Motion
$\sigma_{\mathcal{N}} = 0$	$1.3 \cdot 10^{-10}$	$6.23 \cdot 10^{-15}$
$\sigma_{\mathcal{N}} = 0.1$	$1.8 \cdot 10^{-2}$	$5.08 \cdot 10^{-5}$
$\sigma_{\mathcal{N}} = 0.2$	$3.6 \cdot 10^{-2}$	$9.80 \cdot 10^{-5}$
$\sigma_{\mathcal{N}} = 0.5$	$8.2 \cdot 10^{-2}$	$2.12 \cdot 10^{-4}$
$\sigma_{\mathcal{N}} = 1.0$	$1.8 \cdot 10^{-1}$	$5.06 \cdot 10^{-4}$
$\sigma_{\mathcal{N}} = 2.0$	$3.5 \cdot 10^{-1}$	$9.8 \cdot 10^{-4}$
$\sigma_{\mathcal{N}} = 5.0$	$8.4 \cdot 10^{-1}$	$2.5 \cdot 10^{-3}$

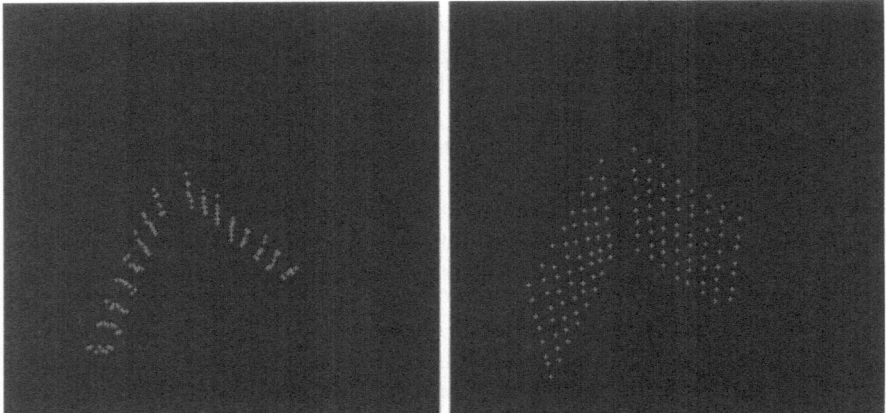

Fig. 6.10. Two perspective views of the 3D reconstruction, simulation with a noise of 1 pixel standard deviation

which are quite accurate values since the reconstruction, obtained with a level of noise of $\sigma_{\mathcal{N}} = 1.0$ is given in Fig. 6.10.

We also have computed the residual retinal error after the estimation of the motion parameters, which standard deviation is written $\sigma_{\mathcal{R}}$. Ideally, i.e. if the model integrates all information about the retinal disparities, except the noise, the standard deviation must correspond to the input noise. If not, we have a bias in the estimation process of amplitude b and we can estimate this bias using the following approximate formula [6.39, 40]: $\sigma_{\mathcal{R}}^2 = \sigma_{\mathcal{N}}^2 + b^2$.

We have obtained:

Noise Level	Standard deviation of the residual error	Estimated bias [pixel]
$\sigma_{\mathcal{N}} = 0$	$\sigma_{\mathcal{R}} = 3 \cdot 10^{-12}$	$b = 3 \cdot 10^{-12}$
$\sigma_{\mathcal{N}} = 0.1$	$\sigma_{\mathcal{R}} = 0.107$	$b = 0.038$
$\sigma_{\mathcal{N}} = 0.2$	$\sigma_{\mathcal{R}} = 0.209$	$b = 0.061$
$\sigma_{\mathcal{N}} = 0.5$	$\sigma_{\mathcal{R}} = 0.526$	$b = 0.163$
$\sigma_{\mathcal{N}} = 1.0$	$\sigma_{\mathcal{R}} = 1.117$	$b = 0.497$
$\sigma_{\mathcal{N}} = 2.0$	$\sigma_{\mathcal{R}} = 2.191$	$b = 0.894$
$\sigma_{\mathcal{N}} = 5.0$	$\sigma_{\mathcal{R}} = 5.277$	$b = 1.687$

which shows that the estimation bias is always small with respect to the noise.

The experimentation with $\sigma_{\mathcal{N}} = 0$ corresponds to a set of "perfect data". The obtained results should 0 ideally and the residual errors correspond to rounding errors of the machine. This also shows that the experimentation program has normally no bug!

We have also simulated the fact we were using line correspondences instead of point correspondences. Lines were defined simply using two points of the calibration grid obtained from the previous algorithm. Since we define lines from two points of the data set and only introduce the normal displacements of these two points in the measurement equations we never consider the line parameters, but only sliding points. We have considered error on the normal displacement of the sliding points and have obtained very similar results, for instance:

Noise Level	Motion from Structure	Structure from Motion
$\sigma_{\mathcal{N}} = 1.0$	$2.7 \cdot 10^{-1}$	$7.12 \cdot 10^{-4}$
$\sigma_{\mathcal{N}} = 2.0$	$6.5 \cdot 10^{-1}$	$1.13 \cdot 10^{-3}$

These results have been obtained considering 128 lines containing the 128 points and we thus have generated 256 normal correspondences between those points. This validates our mechanism of integration of line correspondences.

Precision of the Boot-Strapping Phase. We have studied the precision of the boot-strapping phase, which corresponds to the estimation of a fundamental matrix as in [6.33]. Starting from scratch, the minimization is obtained in about 30 iterations and 1.5 s CPU-time in the same conditions as before. Considering the same error measure as above we have obtained:

Noise Level	Motion Only (Boot-Strapping)
$\sigma_{\mathcal{N}} = 0.1$	$9.82 \cdot 10^{-2}$
$\sigma_{\mathcal{N}} = 0.2$	$9.64 \cdot 10^{-2}$
$\sigma_{\mathcal{N}} = 0.5$	$1.04 \cdot 10^{-1}$
$\sigma_{\mathcal{N}} = 1.0$	$1.07 \cdot 10^{-1}$
$\sigma_{\mathcal{N}} = 2.0$	$3.40 \cdot 10^{-1}$
$\sigma_{\mathcal{N}} = 5.0$	$6.35 \cdot 10^{-1}$

which shows that the precision has the same order of magnitude than what has been obtained by the rest of the module; it is more sensitive to small errors because this estimation is definitely not linear, while the previous one was almost linear as discussed previously. However, we perform here the minimization with respect to 7 parameters instead of 11 and it is known [6.36] that the smaller the number of parameters, the more stable the algorithm. This might explain why this minimization loop is less sensitive to high errors.

Limitations in the Convergence of the Algorithm, Simulation. As a nonlinear algorithm, the proposed module might not converge to the correct solution, if the initial value is too far from what has been expected.

We have observed that we could start from any motion values if the proximity had a relative variation of 20 % with respect to their true values, and similarly we could start from any proximity values if the motion values had a relative variation of 10 % with respect to their true values.

Moreover the algorithm is still converging when both motion and proximity values relative variations were about 50 % of their true values.

This radius of convergence is rather small, and it is then of primary importance to use bootstrapping mechanisms to ensure the convergence of the method.

Precision of the Algorithm, Real Data. We finally have checked our algorithm on the real data and have obtained the following precision:

Real Data	Motion from Structure	Structure from Motion
	$7.39 \cdot 10^{-1}$	$3.16 \cdot 10^{-3}$

which is again quite accurate, although it corresponds to a noise level of about five pixels. Some points have been quite badly located as visible on

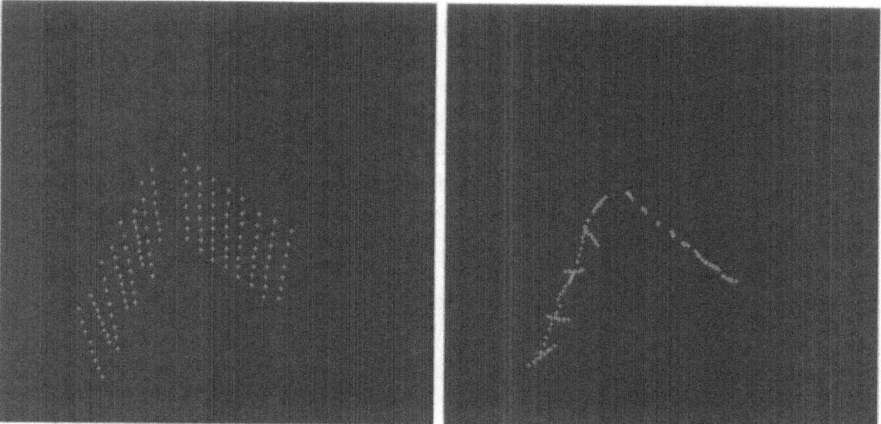

Fig. 6.11. Two perspective views of the 3D reconstruction, real data

the reconstruction given in Fig. 6.11. When we have observed this fact, we were not disappointed because this was an opportunity to check whether the algorithm could deal with important errors, and it did.

6.5 Conclusion

Considering the problem of computing structure and motion, given a set of point and/or line correspondences, in a monocular image sequence, when the camera is **not** calibrated, we have analyzed the underlying algebraic problem and have proposed suitable representations, depending on the chosen geometry. Among these representations the Qs-representation which seems to be the simplest generalization of what is well known in the calibrated case.

We have verified that this representation is in deep relation with what has been recently studied in the field: the Fundamental matrix, the 3D reconstruction up to an affine or projective transform and the problem of auto-calibration, and can be used in the case of line correspondences.

Considering the case of 3 views we have studied in detail the different aspects and possible algorithms of estimation and have developed a geometric interpretation in this case.

At a first glance, the obtained results are quite negative: no possibility of auto-calibration in the general case unless some additional hypotheses are taken into account, apparent increase of the number of constraints between the different parameters involved in this representation when the number of views increases, difficulties to derive complete algorithms for N views and point+line correspondences, etc.

However, using a composite representation based on the Qs-representation it is possible to avoid these difficulties and to estimate the $11 N-4$ parameters related to the retinal motion and the proximity of the points, corresponding to their affine depths.

If we follow this method, we can obtain a very accurate model of the retinal correspondences and the corresponding numerical values are much more accurate than what has been obtained in the calibrated case. This might be due to the fact we use more parameters, and the right ones, so that we do not have to trust the very unstable values of a calibration module.

It is clear that this book does not modify the conclusions obtained by previous researchers who have already analyzed visual data when calibration was not given, but we have tried to extend previous studies in the field in three directions: (1) variable calibration parameters, (2) line and point correspondences, (3) not only 2/3 views but many.

6.6 Application to the Planar Case

In order to emphasis the generality of the approach proposed in this book, let us briefly develop also the planar case, and one example of parametric curves. This might help to illustrate some key ideas developed in this book.

In some situations, we can assume that all points belong to same 3D plane S. We parameterized this plane considering its normal N_j, with $\|N_j\| = 1$ and its distance to the origin d_j, these quantities being taken in frame j. We can write the plane equation as [6.16]:

$$M \in P \quad \Leftrightarrow \quad M^{\mathrm{T}} \cdot N_j = d_j \ . \tag{6.39}$$

Combining this equation with (6.1) and (6.2) leads to

$$Z_i \, m_i = Z_j \, \underbrace{\left(Q_{ij} + s_{ij} \cdot n_j^{\mathrm{T}}\right)}_{H_{ij}} \cdot m_j \quad ; \quad n_j = \frac{A_j^{-1\,\mathrm{T}} \cdot N_j}{d_j} \tag{6.40}$$

as developed in [6.22]. With this relation, we obtain the well-known fact that the retinal displacement of the projection of a planar patch is entirely defined by a collineation, and explicit the relations between this collineation and the Qs-representation. The n_j vector of this paragraph, resumes the information about the structure of the plane and corresponds to the "uncalibrated" plane equation since if m is the projection of a point of the plane P we obviously have $n_j^{\mathrm{T}} m = 1$.

Using this equation we can derive some useful relationships between the collineation and the fundamental matrix. We have

$$\tilde{s}_{ij} \cdot H_{ij} = \tilde{s}_{ij} \cdot Q_{ij} = F_{ij} \tag{6.41}$$

for any collineation H_{ij}, while

$$F_{ij}^{\mathrm{T}} \cdot H_{ij} = F_{ij}^{\mathrm{T}} \cdot Q_{ij} = S_{ij}^{\mathrm{T}} \cdot \tilde{s}_{ij} \cdot S_{ij} = \frac{\tilde{s}_{ji}}{-\det(Q_{ji})}$$

is a skew-symmetric matrix as already used by [6.32], and we also have, using (6.40): $H_{ij} \cdot s_{ji} = - s_{ij}$. And finally using the previous decomposition of a Q-matrix we derive

$$H_{ij} = S_{ij} + s_{ij} \cdot (r_{ij} + n_{ij})^{\mathrm{T}} . \qquad (6.42)$$

This last equation is very important because it clearly shows, that given any collineation corresponding to an unknown plane, we measure $r_{ij} + n_{ij}$ only, and therefore the r_{ij} vector is not measurable in an independent way.

For the plane at infinity, $d_j^{-1} = 0$ and $H_{ij} = Q_{ij}$ as expected. Reciprocally, if Q_{ij} is known, the location of the plane at infinity is known since we can determine given point correspondences, whether they belong to this plane or not. This have been found in (6.3), considering points with huge depths.

Moreover, we can provide a geometric interpretation of the r_{ij} vector, since the quantity $r_{ij} + n_{ij}$ is just equal to r_{ij} for the plane at infinity. Therefore this vector simply characterizes the affine structure of the scene by locating the plane at infinity.

Finally the S-matrix can also be interpreted as follows: because S_{ij} is singular, it is not a collineation, but still a correspondence between the two retinas induced by a plane, which contains one optical center (this explains why the transformation is singular). The image of a point m_j in the frame j is $m_i = \lambda\, s_{ij} \wedge F_{ij} \cdot m_j$ in other words, the intersection of the epipolar associated to m_j with the line $\delta_{s_{ij}}$ of equation $m \in \delta_{s_{ij}} \Leftrightarrow m^{\mathrm{T}} \cdot s_{ij} = 0$, as developed in this book.

Considering now (6.41) we can make the following remark: this equation corresponds to the "motion equation for planar patches" in the sense that we have eliminated the parameters related to the structure and have summarized all information about motion, and independent of the structure of the plane, in this equation.

Now, the Qs-representation being known, it is very easy to recover the uncalibrated structure of the plane since

$$n_j = \frac{\lambda H_{ij}^{\mathrm{T}} \cdot s_{ij} - r_{ij}}{s_{ij}^{\mathrm{T}} \cdot s_{ij}} \quad ; \quad \lambda = \frac{\|\tilde{s}_{ij} \cdot Q_{ij}\|}{\|\tilde{s}_{ij} \cdot H_{ij}\|} . \qquad (6.43)$$

Moreover, combining (6.1) and (6.2) with (6.39) we calculate the evolution of the parameter related to the structure of the plane and easily obtain

$$n_j = \frac{Q_{ij}^{\mathrm{T}} \cdot n_i}{1 - s_{ij}^{\mathrm{T}} \cdot n_i} .$$

Therefore, as for points and lines, the uncalibrated parameters related to the structure of the plane are entirely defined by the Qs-representation.

From these relations and (6.4) we immediately deduce

$$
\begin{aligned}
H_{ii} &= 0 , \\
H_{ji} &= H_{ij}^{-1} , \\
H_{ik} \cdot H_{kj} &= H_{ij}
\end{aligned}
$$

which is not a surprise, since Q-matrices behave like all collineations (see (6.4)).

This digression about the planar case illustrates one key idea of this book: using the Qs-representation, or another, we can derive the equations about (1) motion, (2) structure from motion or (3) evolution of structure parameters as in the calibrated case, but using the "retinal projection" of the Euclidean parameters. We conjecture that this can be done for many other primitives.

This is, for instance, also the case for a planar conic. A planar conic is defined by the intersection of a plane and of a quadric, say:

$$\begin{cases} M^\mathrm{T} \cdot N_i = d_i & \text{plane equation ,} \\ M^\mathrm{T} \cdot C_i \cdot M + d_i^\mathrm{T} \cdot M + e_i = 0 & \text{quadric equation} \end{cases}$$

where the 3×3 matrix C_i, the vector d_i and the scalar e_i are, up to a scale factor, the parameters of a $3D$ quadric. It can be easily shown that the retinal projection of this conic is in homogeneous coordinates:

$$m_i^\mathrm{T} \cdot A_i^{-1\,\mathrm{T}} \cdot \underbrace{\left(C_i + d_i \cdot \frac{N_i}{d_i}^\mathrm{T} + e_i \frac{N_i}{d_i} \cdot \frac{N_i}{d_i}^\mathrm{T} \right) A_i^{-1}}_{\Gamma_i} \cdot m_i = 0 \ .$$

The Γ_i matrix represents, with $n_i = \left(A_i^{-1\,\mathrm{T}} \cdot N_i \right)/d_i$ the "uncalibrated" structure of the conic, the former being the $2D$ information (what is measured in the image) and the latter the $3D$ information (what should be estimated to recover the scene structure). The geometrical interpretation is that Γ_i defines both the retinal image of the conic and a $3D$ cone K_i, which contains the $3D$ conic, this curve being the intersection of the $3D$ cone with the $3D$ plane defined by its projection n_i. Therefore the structure from motion problem is to recover n_i as developed in this paragraph. Moreover, the prediction of Γ_i from one frame to another is given by the following formula:

$$\begin{aligned} \Gamma_j &= Q_{ij}^\mathrm{T} \cdot \Gamma_i \cdot Q_{ij} + Q_{ij}^\mathrm{T} \cdot \Gamma_i \cdot s_{ij} \cdot n_j^\mathrm{T} \\ &+ n_j \cdot s_{ij}^\mathrm{T} \cdot \Gamma_i \cdot Q_{ij} + n_j \cdot s_{ij}^\mathrm{T} \cdot \Gamma_i \cdot s_{ij} \cdot n_j^\mathrm{T} \end{aligned}$$

and thus entirely defined by the Qs-representation, again.

7. Conclusion

In this study we have developed several modules which can be used in a reactive system, in order to perform 3D vision. At the end of this study, it is clear that the capabilities of an active visual system are enhanced when introducing some 3D vision parameters:

Considering the **where to look next problem**, as soon as one can distinguish the effect of 3D rotations and the effect of 3D translations on retinal disparities, it is possible to isolate areas in the image on which it is natural to focus attention. These areas correspond to regions of residual disparity, the rotational disparity being canceled. Rotational disparity can be canceled in several ways, either using inertial cues (Sect. refsection:4.4), or visual motion analysis using vertical estimates (Sect. 4.6), or detecting the collineation of the plane at infinity, i.e. motion of points at the horizon (Sect. 5.2). These motion cues for focus of attention are, in fact, often used implicitly in the literature [7.1, 2] but are not made explicit by the authors, whose paradigms are always limited to some relatively simple scenes. Moreover, it is a reasonable assumption to assume that biological systems of rudimentary vertebrates, capable of such mechanisms of attention focusing (e.g. a frog catching a fly) are based on some non-elaborated mechanisms of motion perception, as observed in the sub-cortical layers of the visual system [7.3]. There is a link with our work here. We – indeed – do not pretend to have a model of such mechanisms here, but that our algorithms have somehow a relatively low complexity, comparable to what such neuronal layers are doing [7.4].

Considering now the problem of **visual target tracking** it has been experimentally shown that there is a benefit in introducing a 3D kinematic model for the target. The reason is very simple: considering usual objects, their 3D rigid motion is often quite regular, and locally of constant velocity or acceleration. The retinal projection of this motion might be much more complex and rather non regular especially in the peripheral part of the visual field, for obvious geometrical reasons. But, of course, a target enters first the peripheral part of the visual field and this is the place where its motion is to be evaluated as it is the case in primates [7.3]. Therefore, and contrary to what is usually considered for biological systems [7.5], or claimed in artificial active vision [7.1, 6, 7] it is a fruitful strategy to explicitly introduce a 3D internal model of the target (Sect. 2.3). It will not only allow to estimate the target kinematic, thus predict its next location and compensate for system delays, but also allow to observe and analyse the target kinematic (Sect. 2.4). Knowing that a target is of constant velocity or has an almost vertical parabolic trajectory might help to classify the kind of observed target as illustrated in Fig. 7.1. In this diagram, we have generalized the idea of estimating the parameters of a visual target in a 3D frame of reference and of considering a set of trivial constraints (some components of the acceleration or velocity are equal to zero) as done in this book and it is clear that some qualitative aspects of the target motion can easily be obtained as done in dynamic vision [7.8]. In addition, it has been demonstrated (Sect. 2.2) that introducing the depth of the target as a parameter of the system allows to control the robotic head in one step, since we can entirely predict the configuration (or the subset of configurations in case of redundancies) of the mount for a fixation on the given 3D location. This includes focus, and can be extended to zoom control. Finally (Sect. 2.4), a rudimentary 3D model allows to perform the fusuion between vergence, zoom and focus cues and to obtain a simple and direct interaction *perception* ↔ *action* since there is a one to one correspondence between the perception of a target location and the head configuration. Let us add a last remark: considering statistical frameworks *à la* Kalman, it is wrong to say that using unreliable estimations of the target depth might de-stabilize the system: if the information on the target depth is unreliable, its information (inverse of the variance) is simply set to zero and the system acts as if it were using 2D information only. There is thus a gain and no loss, except a – relatively small – increase of complexity.

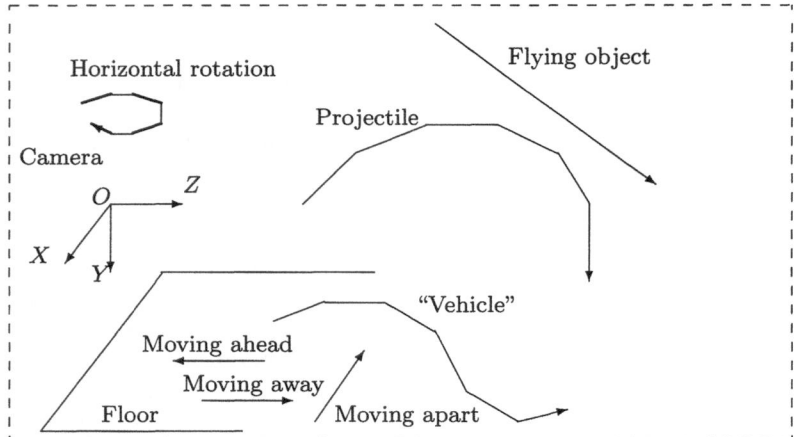

Fig. 7.1. Purposive interpretation of 3D motion. Considering a 3D kinematic model of the target, several attributes can be inferred by simply detecting whether some components of the target velocity/acceleration are zero

One step further, we have demonstrated that **we indeed can perform some 3D vision on a robotic head** (Sect. 2.1). This was not obvious at first and has required to solve a rather painful problem: the calibration of the robotic head [7.9]. Thanks to some accurate knowledge of how intrinsic parameters of a visual sensor with focus and/or zoom capabilities vary [7.10] it has been possible to use a restrained version of these functional models for our mount. However it is clear that the success of this experiment is deeply related to the fact that our robotic head is somehow "simple". More sophisticated devices have not yet been calibrated [7.1, 11] or only partially [7.12]. The solution to deal with such complex systems is **to be able to auto-calibrate the intrinsic and extrinsic parameters of a visual sensor from the environment**. In this study (Chap. 3), and contrary to other recent developments in the field [7.13–15], we have not only obtained the intrinsic but also the extrinsic parameters of the visual system. Moreover, because we have intentionally made use of a specific method, we have obtained rather good performances compared to more general methods [7.13, 16], while the uncertainty of the estimation can be easily determined in our paradigm (Sect. 3.5). Finally we also can very easily decide which head movement to perform in order to calibrate a given subset of the calibration parameters (Sect. 3.6), i.e. pan rotation to estimate horizontal parameters and pitch rotation to estimate vertical parameters.

If we want to analyse 3D visual data in the field of active vision, it is clear that we do not only want a "depth map" but to analyse the visual scene in a somehow purposive fashion [7.17–19] or more specifically **with some ecological concepts** in the sense of [7.20]. This aspect of active visual perception is only touched here, but two ideas have been introduced. On the first hand, **the depth is not considered as an isotropic quantity** whereas we make the distinction between (a) tokens at the horizon, i.e. with negligible depth, (b) tokens in an intermediate space for which the orientation can not be perceived because they are too far away and (c) tokens close to the observer and really perceived in 3D. These ideas come to light from the particular analysis of the structure and motion proposed in this book (Sect. 4.6) and correspond to what is usually referred as the grasping space, environment space and horizon in human visual perception [7.21]. In the grasping space, distances have to be measured precisely in order to realize sensory-motor tasks, since the task of taking an of objects requires such an information. In the environment space, or intermediate space, it is only required to approximate the depths of objects in order to decide for potential alarms, say prey or predator. On the other hand, because **we have introduced the notion of vertical in the heart of our representation** (Sect. 4.6), the perception of the environment is not isotropic but always attached to the orientation with respect to the vertical. Therefore, it is clear [7.20] that several rudimentary notions of floor (horizontal surface at a given height), wall (as being vertical surface) and left and right arise and can easily be formalized using some geometric constraints.

This is illustrated in Fig. 7.2. As a conclusion on this topic we can say that the active vision paradigm has helped to implement some original aspects of visual perception which can not be taken into account if we do not consider that *the visual system is embedded in its environment.*

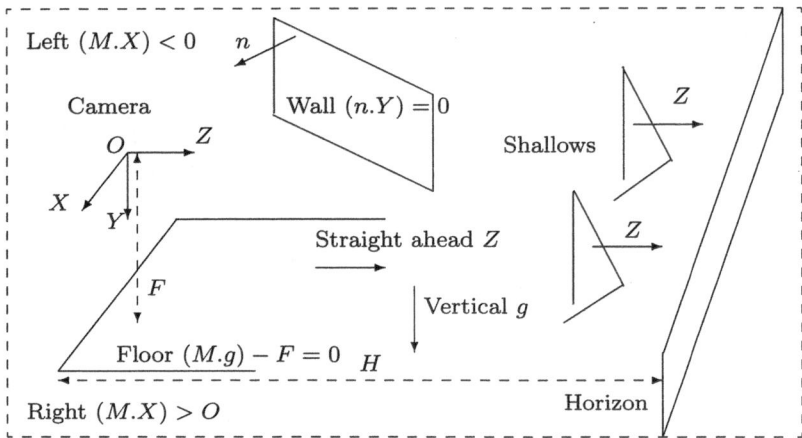

Fig. 7.2. Purposive interpretation of depth and orientation. Considering the vertical cue and some constraints on the depth, and the object relative position with respect to planes, it is possible to complete the depth recovery by some attributes on the object characteristics

The problem of visual perception can not be decoupled from the problem of **inertial and visual data fusion**. This is an evidence for biological visual systems [7.3, 4], but is surprisingly often ignored by the computer science community despite several publications in this direction [7.22–26]. As far as active vision is concerned, we have reviewed how inertial cues can be embedded in a visual system and what can be easily obtained: (1) rotational velocity estimation, (2) vertical direction perception, (3) elimination of the monocular scale-factor undetermination, etc. The main idea here is that the fusion of these two modalities does not only require the fusion of data with different geometries or dynamics but also to somehow change the internal visual representation as proposed in Sect. 4.6.

As far as 3D vision is concerned, the interaction between active vision and 3D visual perception is not limited to Euclidian 3D vision but **includes either affine or projective 3D models**. This corresponds to the general outcomes of the last chapter. However, these ideas have not been introduced after some complex and somehow tedious geometrical developments, but simply using algebraic derivations. Affine geometry is sufficient to calibrate a visual system with constant intrinsic parameters (Sect. 5.2) and leads to linear equations. As a consequence, the analysis of the measurability of these parameters, has been easily performed. Affine geometry again, is sufficient to derive criteria to detect planar objects in motion. Moreover, some pieces of information about the "uncalibrated structure of the object" i.e. the projection of its 3D parameters can be computed and predicted from one frame to another without any calibration. These quantities can be easily related to some ratios or cross-rations of Euclidian distances as made explicit here. By restraining our model of the scene to the simple case of piece-wise planar patches we have been able to not only derive theoretical ideas but to also implement in a somewhat robust way retinal motion estimation (Sect. 5.3), with a precision sufficient to infer some 3D parameters of the scene (Sect. 5.4). This restriction is related to the fact that a reasonable symbolic description of the scene in the field of active vision is better done using a rather "dense" representation considering regions of homogeneous intensity (Sect. 2.3) or of coherent retinal motion (Sect. 5.3) than as parse representation (edges, corners) with which we might lose some data, may be potential alarms, or objects requiring some attention. Here we "see everything" although without a big precision. But the same ideas can be applied to points or lines [7.27], or conics [7.28].

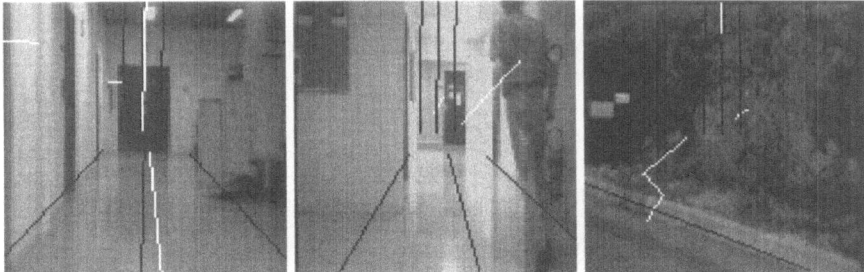

Fig. 7.3. Examples of autonomous navigation, using 3D reactive vision. When a 3D representation of the visual scene is used, it is very easy to detect the left and/or right borders of a road or corridor, both indoors and outdoors. Using this information the control of a mobile robot is straightforward

We finally would like to leave the reader with a last example showing how 3D vision might dramatically simplify a reactive behavior. Let us consider the visual system of a mobile robot. Its task is not only to execute a predefined trajectory, but also compensate for accumulative errors and uncertainties in its initial map, i.e. stabilize the mobile platform on its trajectory. Shall we use 2D or 3D vision? Several clever algorithms exist which analyse 2D images. But they are very often rather complex and can not be used in any environment. On the contrary, we make use of a real-time 3D stereo system [7.26] and which simply detects the left and/or right border of the corridor or road, using a threshold on the scene height. Using about 20 lines of code [7.29] the system resets the robot in the middle of its track. This works, as shown in Fig. 7.3

and as detailed in [7.26]. This problem is simplified just because we were in 3D. The 3D model of the middle of a track is a very simple feature, whereas its 2D projection might be much more complicated to identify. Therefore, a mobile robot better uses a 3D visual system to navigate.

We are confident that this is not an exception, whereas many other active perceptual tasks are to be performed in 3D.

References

Chapter 1

1.1 J.L. Plotter. The Staran Architecture and Its Application to Image Processing and Pattern Recognition Algorithm, 1978.

1.2 G. Granlund. *GOP: A Fast and Flexible Processor for Image Analysis.* Academic Press, 1981.

1.3 E. Appiani, G. Barbagelata, F. Cavaganro, B. Conterno, and R. Manara. An Industry Developped Hierarchical Multiprocessor for Very High Performance Signal Processing Applications. In: *Proc. of the 1st International Conference on Supercomputing Systems*, 1985.

1.4 H. Inoue and H. Mizoguchi. A Flexible Multi-Window Vision-System for Robots. In: *Proc. of the 2nd International Symposium on Robotics Research*, pp. 95–102, 1985.

1.5 J.G. Harris and A.M. Flynn. Object Recognition Using the cm's Router. In: *Proc. IEEE CVPR*, 1986.

1.6 D.H. Schaefer, P. Ho, J. Boyd, and C. Vallejos. The GAM Pyramid. In: *Parallel Computer Vision*, ed. by L. Uhr, Chap. 2. Academic Press, 1989.

1.7 C.M. Brown. Parallel Vision on the Butterfly Computer. In: *Proc. of the 3rd International Conference on Supercomputing Systems*, 1988.

1.8 B.S. Sunwoo, M.H. Baroody, and J.K. Aggarwal. A Parallel Algorithm for Region Labeling. In: *Proc. of the 1987 Workshop on Computer Architecture for Pattern Analysis and Machine Intelligence: CAPAMI '87*, 1987.

1.9 O.D. Faugeras, R. Deriche, N. Ayache, F. Lustman, and E. Giuliano. Depth and Motion Analysis: The Machine Being Developed Within Esprit Project 940. In: *Proc. of the IAPR Workshop on Computer Vision (Special Hardware and Industrial Applications), Tokyo, Japan* pp. 35–44. Institute of Industrial Science, University of Tokyo, October 1988.

1.10 R. Vaillant, R. Deriche, and O.D. Faugeras. 3D Vision on the Parallel Machine CAPITAN. In: *International Workshop on Industrial Application of Machine Intelligence and Vision*, April 1989.

1.11 E. Barton. Data Concurrency on the Meiko Computing Surface. In: *Parallel Processing for Computer Vision and Display*, ed. by P.M. Dew, R.A. Earnshaw, and T.R. Heywood, Chap. 29. Addison-Wesley, 1989.

1.12 J. Deutch, P.C. Maulik, R. Mosur, H. Printz, H. Ribas, J. Senko, P.S. Tseng, J.A. Webb, and I.C. Wu. Performance of Warp on the DARPA Image Understanding Benchmarks. In: *Parallel Processing for Computer Vision and Display*, ed. by P.M. Dew, R.A. Earnshaw, and T.R. Heywood, Chap. 8. Addison-Wesley, 1989.

1.13 H. Wang. Corner Detection for 3-D Vision Using Array Processors. In: *Proc. BARNAIMAGE 91*, 1991.

1.14 H.H. Liu and T.Y. Young. VLSI Algorithms and Structures for Consistent Labeling. In: *Proc. of the 1987 Workshop on Computer Architecture for Pattern Analysis and Machine Intelligence: CAPAMI '87*, 1987.

1.15 K. Chen, P.E. Danielsson, P. Emanuellson, and P. Ingelhag. Single-Chip High Speed Computation of Optical Flow. *Proc. MVA '90, Iapr Workshop on Machine Vision Applications*, pp. 331–335, November 1990.

1.16 E.D. Dickmanns. System Architecture for Road-Vehicles Capable of Vision. *PROMETHEUS PRO-ART Workshop On Intelligent Co-Pilot*, pp. 19–29, December 1991.

1.17 S. Rygol, M. Pollard, and C. Brown. Marvin and Tina: A Multiprocessor 3-D Vision System. *Concurrency: Practice and Experience*, 3(4):333–356, August 1991.

1.18 M.J. Swain and M. Stricker. Promising Directions in Active Vision. Technical Report CS 91-27, University of Chicago, IL, 1991.

1.19 K.H. Cornog. Smooth Pursuit and Fixation for Robot Vision. MSc Thesis, 1985.

1.20 E.P. Kroktov. Exploratory Visual Sensing for Determining Spatial Layout with an Agile Stereo Camera System. PhD Thesis, 1987.

1.21 C.M. Brown. Gaze Controls with Interactions and Delays. Technical Report, OUEL: 1770/89, 1989.

1.22 F. Chaumette. *La Commande Référenciée Vision*. PhD Thesis, University of Rennes, Dept. of Comp. Science, 1990.

1.23 K. Pahlavan, J.-O. Ekhlund, and T. Uhlin. Integrating Primary Ocular Processes. In: *2nd European Conference on Computer Vision*, pp. 526–541. Springer-Verlag, 1992.

1.24 N.P. Papanikolopoulos, B. Nelson, and P.K. Khosla. Full 3D Tracking Using the Controlled Active Vision Paradigm. In: *The 7th IEEE Symposium on Intelligent Control, Glasgow*, August 1992.

1.25 T. Viéville, J.O. Ekhlund, K. Pahlavan, and T. Uhlin. An Example of Artificial Oculomotor Behavior. In: *Seventh IEEE Symposium on Intelligent Control, Glasgow*, ed. by T. Henderson, pp. 348–353. IEEE Computer Society Press, 1992.

1.26 T.J. Olsen and R.D. Potter. Real Time Vergence Control. Technical Report, University of Rochester, Comp. Sci.: 264, 1988.

1.27 J.M. Lavest, G. Rives, and M. Dhome. 3D Reconstruction by Zooming. In: *Intelligent Autonomous System, Pittsburg*, 1993.

1.28 K. Pahlavan, T. Uhlin, and J.-O. Ekhlund. Dynamic Fixation. In: *4th International Conference on Computer Vision*, pp. 412–419. IEEE Society, 1993.

1.29 B. Espiau and P. Rives. Closed-Loop Recursive Estimation of 3D Features for a Mobile Vision System. In: *IEEE Conference on Robotics and Automation, Raleigh*, 1987.

1.30 F. Chaumette and S. Boukir. Structure from Motion Using an Active Vision Paradigm. In: *11th International Conference on Pattern Recognition, The Hague, Netherlands*, 1991.

1.31 Q.-T. Luong and O.D. Faugeras. Active Head Movements Help Solve Stereo Correspondence? In: *Proc. ECAI 92*, pp. 800–802, 1992.

1.32 C.M. Brown. The Rochester Robot. Technical Report, University of Rochester, Comp. Sci.: 257, 1988.

1.33 T. Viéville. Real Time Gaze Control: Architecture for Sensing Behaviours. In: *The 1991 Stockholm Workshop on Computational Vision, Rosenon, Sweden*, ed. by J.-O. Eklundh. Royal Institute of Technology, Stockholm, Sweden, 1991.

1.34 C.W. Urquhart and J.P. Siebert. Development of a Precision Active Stereo System. In: *Proc. International Symposium on Intelligent Control, Glasgow,* 1992.

1.35 H.P. Trivedi. Semi-analytic Method for Estimating Stereo Camera Geometry from Matched Points. *Image and Vision Computing,* 9, 1991.

1.36 O.D. Faugeras, Q.T. Luong, and S. Maybank. Camera Self-Calibration: Theory and Experiment. In: *2nd European Conference on Computer Vision, Genoa,* 1992.

1.37 N.A. Thacker. On-line Calibration of a 4-dof Robot Head for Stereo Vision. In: *British Machine Vision Association Meeting on Active Vision, London,* 1992.

1.38 T. Viéville. Autocalibration of Visual Sensor Parameters on a Robotic Head. *Image and Vision Computing,* 12, 1994.

1.39 J.L.S. Mundy and A. Zisserman. *Geometric Invariance in Computer Vision.* MIT Press, Boston, MA, 1992.

1.40 O.D. Faugeras. *Three-Dimensional Computer Vision: A Geometric Viewpoint.* MIT Press, Boston, MA; 1993.

1.41 E. Francois and P. Bouthemy. Multiframe-Based Identification of Mobile Components of a Scene with a Moving Camera. In: *Conference on Computer Vision and Pattern Recognition, Hawai,* pp. 166–172. IEEE Computer Society Press, Alamitos, CA, 1991.

1.42 D.H. Ballard and C.M. Brown. *Computer Vision.* PrenticeHall Inc., New Jersey, USA, 1982.

1.43 J.J. Koenderink and A.A.J. van Doorn. Geometry of Binocular Vision and a Model of Steropsis. *Biological Cybernetics,* 21:29–35, 1976.

1.44 W.E.L. Grimson. *From Images to Surfaces.* MIT Press, Cambridge, MA, 1986.

1.45 J.E.W. Mayhew. The Interpretation of Stereo-Disparity Information: The Computation of Surface Orientation and Depth. *Perception,* 11:387–403, 1985. And see the appendix by H.C. Longuet-Higgins, pp. 405–407.

1.46 H.C. Longuet-Higgins. Appendix to Paper by John Mayhew Entitled: 'The Interpretation of Stereo-Disparity Information: The Computation of Surface Orientation and Depth'. *Perception,* 11(4):405–407, 1985.

1.47 B.F. Buxton, H. Buxton, and A. Kashko. Optimization, Regularization, and Simulated Annealing in Low-Level Computer Vision. In: *Parallel Architectures and Computer Vision.* Clarendon Press, Oxford, 1988.

1.48 E.H. Thompson. A Rational Algebraic Formulation of the Problem of Relative Orientation. *J. Photogrammetric Record,* 3(14):152–159, 1959.

1.49 H.C. Longuet-Higgins. A Computer Algorithm for Reconstructing a Scene from Two Projections. *Nature,* 293:133–135, 1981.

1.50 R.Y. Tsai and T.S. Huang. Three Dimensional Motion Estimation from Image Space Shifts. *J. IEEE Trans. on ASSP,* 1981.

1.51 A.N. Ayache and A.F. Lustman. Fast and Reliable Passive Trinocular Stereovision. In: *Proc. of the IEEE 1st International Conference in Computer Vision,* ed. by J.M. Brady and A. Rosenfeld, pp. 422–427. IEEE Computer Society Press, Alamitos, CA, 1987.

1.52 S.B. Pollard, J.E.W. Mayhew, and J.P. Frisby. Pmf – a Stereo Correspondence Algorithm Using a Disparity Gradient Limit. *Perception,* 14(4):449–470, 1985.

1.53 B.K.P. Horn. *Robot Vision.* MIT Press, Cambridge, MA, 1986.

1.54 G. Toscani and O.D. Faugeras. Camera Calibration for 3D Computer Vision. In: *Proc. of the International Workshop on Machine Intelligence, Tokyo,* February 1987.

1.55 R.Y. Tsai. An Efficient and Accurate Calibration Technique for 3D Machine Vision. In: *IEEE Proc. CVPR '86, Miami Beach, FL*, pp. 364–374, June 1986.

1.56 B. Caprile and V. Torre. Using Vanishing Points for Camera Calibration and Motion Recovery. Int. Report, Dept. of Physics, University of Genoa, 1987.

1.57 A. Izaguirre et al. A New Development in Camera Calibration – Calibrating a Pair of Mobile Cameras. *IEEE Conference on Robotics and Automation*, pp. 74–79, 1985.

1.58 A.L. Abbott and N. Ahuja. Surface Reconstruction by Dynamic Integration of Focus, Camera Vergence and Stereo. *2nd International Conference on Computer Vision*, p. 532, 1988.

1.59 J.J. Clark and N.J. Ferrier. Modal Control of an Attentive Vision System. *IEEE International Conference on Computer Vision*, 1988.

1.60 A.L. Abbott and N. Ahuja. Active Surface Reconstruction by Integrating Focus, Vergence Stereo and Camera Calibration. In: *Proc. of the 3rd International Conference on Computer Vision, Osaka*, 1990.

1.61 R. Boudarel, J. Delmas, and P. Guichet. *Commande Optimale des Processus.* Dunod, Paris, 1968. Vol. 2, 3, 4.

1.62 S. Samson, M. le Borgne, and B. Espiau. *Robot Control: The Task-Function Approach.* Oxford University Press, 1990.

1.63 B. Espiau, F. Chaumette, and P. Rives. Une Nouvelle Approche de la Relation Vision-Commande en Robotique. Technical Report RR-1172, INRIA, Sophia, France, 1988.

1.64 B. Espiau, F. Chaumette, and P. Rives. A New Approach to Visual Servoing in Robotics. *IEEE Trans. on Robotics and Automation*, 1991. In press.

1.65 P. Rives. Dynamic Vision : Theoretical Capabilities and Practical Problems. In: *NATO Workshop on "Kinematic and Dynamic Issues in Sensor Based Control", Italy*, October 1990.

1.66 A. Berthoz and G.M. Jones. *Adaptive Mechanism in Gaze Control.* Elsevier, Amsterdam, 1985.

1.67 R. Grupen and T.C. Henderson. Autochtonous Behaviors – Mapping Perception to Action. In: *Traditional and Non-Traditional Robotic Sensors*, ed. by T. Henderson. Springer-Verlag, Berlin, September 1989.

1.68 R. Lumia. Sensor-Based Robot Control Requirement for Space. In: *Traditional and Non-Traditional Robotic Sensors*, ed. by T. Henderson. Springer-Verlag, Berlin, September 1989.

1.69 T. Viéville and P. Sander. Using Pseudo Kalman-Filters in the Presence of Constraints. Technical Report RR-1669, INRIA, Sophia, France, 1992.

1.70 A.S. Santos and F. Chaumette. Reactive Tracking by Visual Servoing. Technical Report PI-683, IRISA, Rennes, France, 1992.

1.71 T. Viéville, E. Clergue, R. Enciso, and H. Mathieu. Experimentating with 3D Vision on a Robotic Head. *Robotics and Autonomous Systems*, 14(1), 1995.

1.72 B. Bascle and R. Deriche. Features Extraction Using Parametric Snakes. In: *Proc. 11th IAPR*, vol. 3, pp. 659–662, August 30–September 3, 1992.

1.73 B. Bascle and R. Deriche. Stereo Matching, Reconstruction, and Refinement of 3D Curves Using Deformable Contours. In: *4th International. Conference on Computer Vision, Berlin*, ed. by H.H. Nagel. IEEE Computer Society Press, Los Alamitos, CA, 1993.

1.74 G. Chow. On Using the md96 Board for Reactive Vision. Technical report, Université de Nice, September 1993. Rapport de Stage de DEA.

1.75 R. Enciso, T. Viéville, and O. Faugeras. Approximation du Changement de Focale et de Mise au Point par une Transformation Affine à Trois Paramètres. Technical Report 2071, INRIA, 1993.

1.76 S. Maybank and O.D. Faugeras. A Theory of Self-Calibration of a Moving Camera. *The International J. of Computer Vision*, 8, 1992.

1.77 Q.-T. Luong and O.D. Faugeras. Self-Calibration of a Camera Using Multiples Images. In: *Proc. 11th ICPR*, pp. 9–12, 1992.

1.78 Q.T. Luong and O.D. Faugeras. Determining the Fundamental Matrix with Planes: Instability and New Algorithms. In: *IEEE Proc. CVPR '93, New-York, June*, pp. 194–199, 1993.

1.79 T. Viéville, P.E.D.S. Facao, and E. Clergue. Computation of Ego-Motion Using the Vertical Cue. *Machine Vision and Applications*, 1994. To appear.

1.80 T. Viéville. Autocalibration of Visual Sensor Parameters on a Robotic Head. In: *Communication at LIFIA, France*, 1992.

1.81 T. Viéville. Vision Modules for Active Vision. Technical Report, Université de Nice, 1993. Hand-Book of the Active Vision Course, INRIA, Sophia, France.

1.82 T. Viéville, P.E.D.S. Facao, and E. Clergue. Building a Depth and Kinematic 3D-Map from Visual and Inertial Sensors Using the Vertical Cue. In: *4th International. Conference on Computer Vision, Berlin*, ed. by H.H. Nagel. IEEE Computer Society Press, Los Alamitos, CA, 1993.

1.83 B. Giai-Checa and T. Viéville. 3D-Vision for Active Visual Loops Using Locally Rectilinear Edges. In: *7th IEEE Symposium on Intelligent Control, Glasgow*, ed. by T. Henderson, pp. 341–347. IEEE Computer Society Press, Alamitos, CA, 1992.

1.84 O.D. Faugeras, B. Hotz, H. Mathieu, T. Viéville, Z. Zhang, P. Fua, E. Théron, L. Moll, G. Berry, J. Vuillemin, P. Bertin, and C. Proy. Real Time Correlation-Based Stereo: Algorithm, Implementations, and Applications. Technical Report 2013, INRIA, 1993.

1.85 B. Giai-Checa, P. Bouthemy, and T. Viéville. Detection of Moving Objects. Technical Report RR-1906, INRIA, Sophia, France, 1993.

1.86 N. Navab and Z. Zhang. A Stereo and Motion Cooperation Approach to Multiple Objects Motion Problems. In: *Proc. of the 2nd Singapore International Conference on Image Processing ICIP*, pp. 518–522, Singapore, September 1992.

1.87 E. Clergue. Méthodes de Reconstruction Denses pour la Vision Active. Technical Report, Université de Nice, September 1993. Rapport de Stage de DEA.

1.88 P. Bouthemy, F. Chaumette, O.D. Faugeras, B. Giai-Checa, Q.T. Luong, L. Robert, A. Santos, and T. Viéville. Real-Time Gaze Control, High-Level Modules for Active Vision. Technical Report, INRIA, Sophia, France, 1993.

1.89 D.W. Murray, P.F. MacLauchlan, I.D. Reid, and P.M. Sharkey. Reactions to Peripheral Image Motion Using a Head/Eye Platform. In: *4th International Conference on Computer Vision*, pp. 403–411. IEEE Society, 1993.

1.90 R.J. Leight and D.S. Zee. *The Neurophysiology of Eye Movement*. Davis Cie, Philadelphia, PA, 1984.

1.91 T. Viéville and O.D. Faugeras. Computation of Inertial Information on a Robot. In: *5th International Symposium on Robotics Research*, ed. by H. Miura and S. Arimoto, pp. 57–65. MIT-Press, 1989.

1.92 O.D. Faugeras, B. Hotz, H. Mathieu, T. Viéville, Z. Zhang, P. Fua, E. Théron, L. Moll, G. Berry, J. Vuillemin, P. Bertin, and C. Proy. Real Time Correlation-Based Stereo: Algorithm, Implementations and Applications. *International J. of Computer Vision*, 1994. In press.

Chapter 2

2.1 T. Viéville. Real Time Gaze Control: Architecture for Sensing Behaviours. In: *The 1991 Stockholm Workshop on Computational Vision, Rosenon, Sweden*, ed. by J.-O. Eklundh. Royal Institute of Technology, Stockholm, Sweden, 1991.

2.2 O.D. Faugeras, R. Deriche, N. Ayache, F. Lustman, and E. Giuliano. Depth and Motion Analysis: The Machine Being Developed Within Esprit Project 940. In: *Proc. of the IAPR Workshop on Computer Vision (Special Hardware and Industrial Applications), Tokyo, Japan*, pp. 35–44. Institute of Industrial Science, University of Tokyo, October 1988.

2.3 T. Viéville and O.D. Faugeras. Computation of Inertial Information on a Robot. In: *5th International Symposium on Robotics Research*, ed. by H. Miura and S. Arimoto, pp. 57–65. MIT-Press, 1989.

2.4 G. Toscani and O.D. Faugeras. Camera Calibration for 3D Computer Vision. In: *Proc. of the International Workshop on Machine Intelligence, Tokyo*, February 1987.

2.5 R.Y. Tsai. An Efficient and Accurate Calibration Technique for 3D Machine Vision. In: *IEEE Proc. CVPR '86, Miami Beach, FL, June*, pp. 364–374, 1986.

2.6 J.M. Lavest, G. Rives, and M. Dhome. 3D Reconstruction by Zooming. In: *Intelligent Autonomous System, Pittsburg*, 1993.

2.7 R. Enciso, T. Viéville, and O. Faugeras. Approximation du Changement de Focale et de Mise au Point par une Transformation Affine à Trois Paramètres. Technical Report 2071, INRIA, 1993.

2.8 T. Viéville, P.E.D.S. Facao, and E. Clergue. Building a Depth and Kinematic 3D-Map from Visual and Inertial Sensors Using the Vertical Cue. In: *4th International Conference on Computer Vision, Berlin*, ed. by H.H. Nagel. IEEE Computer Society Press, Los Alamitos, CA, 1993.

2.9 L. Robert. *Perception Stéréoscopique de Courbes et de Surfaces Tridimensionnelles, Application à la Robotique Mobile*. PhD thesis, Ecole Polytechnique, Palaiseau, France, 1992.

2.10 E. Francois and P. Bouthemy. Derivation of Qualitative Information in Motion Analysis. *Image and Vision Computing*, 8, 1990.

2.11 H.F. Durrant-Whyte. *Integration, Coordination, and Control of Multi-Sensor Robot Systems*. Kluwer Academic Publishers, 1988.

2.12 J.Q. Fang and T.S. Huang. Some Experiments on Estimating the 3-D Motion Parameters of a Rigid Body from Two Consecutive Image Frames. *IEEE Trans. on Pattern Analysis and Machine Intelligence*, 6:547–554, 1984.

2.13 J.K. Aggarwal and W.N. Martin. *Analyzing Dynamic Scenes Containing Multiple Moving Objects*, pp. 355–380. Springer-Verlag, Berlin, 1991.

2.14 T. Viéville, S. Ron, and J. Droulez. Two Dimensional Saccadic and Smooth Pursuit Response to an Extrafoveal Smooth Movement. In: *Proc. of the 3rd European Conference on Eye Movements*, ed. by A. Levy-Shoen and K. O'Reagan. 1986.

2.15 Y.B. Shalom and T.E. Fortmann. *Tracking and Data Association*. Academic-Press, Boston, MA, 1988.

2.16 N. Ayache. *Artificial Vision for Mobile Robots*. MIT Press, Cambridge, MA, 1989.

2.17 R. Deriche and O.D. Faugeras. Tracking Line Segments. In: *Proc. of the 1st European Conference on Computer Vision, Antibes*, pp. 259–269. Springer-Verlag, Berlin, 1990.

2.18 T. Viéville and P. Sander. Using Pseudo Kalman-Filters in the Presence of Constraints. Technical Report RR-1669, INRIA, Sophia, France, 1992.

2.19 R.C.K. Lee. *Optimal Estimation, Identification, and Control.* MIT Press, Cambridge, MA, 1964.

2.20 D.W. Murray, P.F. MacLauchlan, I.D. Reid, and P.M. Sharkey. Reactions to Peripheral Image Motion Using a Head/Eye Platform. In: *4th International Conference on Computer Vision,* pp. 403–411. IEEE Society, 1993.

2.21 T. Viéville, C. Zeller, and L. Robert. Using Collineations to Compute Motion and Structure in an Uncalibrated Image Sequence. *International J. of Computer Vision,* 1995. To appear.

2.22 J. Fairfield. Toboggan Contrast Enhancement. In: *Applications of Artificial Intelligence, Machine Vision and Robotics,* vol. 1708. Proc. of S.P.I.E., 1990.

2.23 K. Pahlavan, T. Uhlin, and J.-O. Ekhlund. Dynamic Fixation. In: *4th International Conference on Computer Vision,* pp. 412–419. IEEE Society, 1993.

2.24 E. Clergue. Méthodes de Reconstruction Denses pour la Vision Active. Technical Report, Université de Nice, September 1993. Rapport de Stage de DEA.

2.25 T. Voorhees and H. Poggio. Detecting Textons and Texture Boundaries in Natural Images. In: *Proc. of the 1st International Conference on Computer Vision, London,* pp. 250–258, June 1987.

2.26 R.O. Duda and P.E. Hart. *Pattern Classification and Scene Analysis.* Wiley Interscience, New-York, 1973.

2.27 S. Das and N. Ahuja. A Comparative Study of Stereo, Vergence and Focus as Depth Cues for Active Vision. In: *IEEE Proc. CVPR '93, New-York, June,* pp. 194–199, 1993.

2.28 O.D. Faugeras, B. Hotz, H. Mathieu, T. Viéville, Z. Zhang, P. Fua, E. Théron, L. Moll, G. Berry, J. Vuillemin, P. Bertin, and C. Proy. Real Time Correlation-Based Stereo: Algorithm, Implementations, and Applications. *International J. of Computer Vision,* 1994. In press.

2.29 O.D. Faugeras. *Three-Dimensional Computer Vision: A Geometric Viewpoint.* MIT Press, Boston, MA, 1993.

2.30 M.J. Anderson. Range from Out-of-Focus Blur. Technical Report LIU–TEK–LIC–1992:17, Linköping University, 1992.

2.31 Y. Xiong and S.A. Shafer. Depth from Focusing and Defocusing. In: *IEEE Proc. CVPR '93, New-York, June,* pp. 68–73, 1993.

2.32 F. Chaumette and A.S. Santos. Tracking a Moving Object by Visual Servoing. In: *Proc. of the 12th World Congress IFAC, Sydney, Autralia,* 1993.

2.33 J.E.W. Mayhew, Y. Zheng, and S.A. Billings. Layered Architecture for the Control of Micro Saccadic Tracking of a Stereo Camera Head. Technical Report AIVRU 72, University of Sheffield, 1992.

2.34 E. Francois and P. Bouthemy. Multiframe-Based Identification of Mobile Components of a Scene with a Moving Camera. In: *Conference on Computer Vision and Pattern Recognition, Hawai,* pp. 166–172. IEEE Computer Society Press, Alamitos, CA, 1991.

Chapter 3

3.1 T. Viéville. Real Time Gaze Control: Architecture for Sensing Behaviours. In: *The 1991 Stockholm Workshop on Computational Vision, Rosenon, Sweden,* ed. by J.-O. Eklundh. Royal Institute of Technology, Stockholm, Sweden, 1991.

3.2 J.-O. Ekhlund, K. Pahlavan, and T. Uhlin. Presentation of the *k*th Robotic Head. In: *5th Scandinavian Workshop on Computational Vision, Rosenon, Sweden,* 1991.

3.3 G. Toscani and O.D. Faugeras. Camera Calibration for 3D Computer Vision. In: *Proc. of the International Workshop on Machine Intelligence, Tokyo*, February 1987.

3.4 R.Y. Tsai. An Efficient and Accurate Calibration Technique for 3D Machine Vision. In: *IEEE Proc. CVPR '86, Miami Beach, FL, June*, pp. 364–374, 1986.

3.5 D.C. Brown. Close-Range Camera Calibration. *Photogrammetric Engineering*, 37, 1971.

3.6 R.Y. Tsai. Synopsis of Recent Progress on Camera Calibration for 3D Machine Vision. *Robotics Review*, 1:147–159, 1989.

3.7 O.D. Faugeras, Q.T. Luong, and S. Maybank. Camera Self-Calibration: Theory and Experiment. In: *2nd European Conference on Computer Vision*, Genoa, 1992.

3.8 S. Maybank and O.D. Faugeras. A Theory of Self-Calibration of a Moving Camera. *The International J. of Computer Vision*, 8, 1992.

3.9 H.P. Trivedi. Semi-analytic Method for Estimating Stereo Camera Geometry from Matched Points. *Image and Vision Computing*, 9, 1991.

3.10 N.A. Thacker. On-line Calibration of a 4-dof Robot Head for Stereo Vision. In: *British Machine Vision Association Meeting on Active Vision*, London, 1992.

3.11 G. Toscani, R. Vaillant, R. Deriche, and O.D. Faugeras. Stereovision Camera Calibration Using the Environment. In: *Proc. of the 6th Scandinavian Conference on Image Analysis*, June 1989.

3.12 E. Francois and P. Bouthemy. Multiframe-Based Identification of Mobile Components of a Scene with a Moving Camera. In: *Conference on Computer Vision and Pattern Recognition, Hawai*, pp. 166–172. IEEE Computer Society Press, Alamitos, CA, 1991.

3.13 A. Rognone, M. Campani, and A. Verri. Identifying Multiple Motions from Optical Flow. In: *2nd European Conference on Computer Vision*, pp. 258–268. Springer-Verlag, 1992.

3.14 B. Giai-Checa, P. Bouthemy, and T. Viéville. Detection of Moving Objects. Technical Report RR-1906, INRIA, Sophia, France, 1993.

3.15 N. Navab and Z. Zhang. From Multiple Objects Motion Analysis to Behavior-Based Object Recognition. In: *Proc. ECAI 92*, pp. 790–794, Vienna, Austria, August 1992.

3.16 T. Viéville. Estimation of 3D-Motion and Structure from Tracking 2D-Lines in a Sequence of Images. In: *Proc. of the 1st European Conference on Computer Vision, Antibes*, pp. 281–292. Springer-Verlag, Berlin, 1990.

3.17 R. Deriche and O.D. Faugeras. Tracking Line Segments. In: *Proc. of the 1st European Conference on Computer Vision, Antibes*, pp. 259–269. Springer-Verlag, Berlin, 1990.

3.18 M.J. Stephens, R.J. Blisset, D. Charnley, E.P. Sparks, and J.M. Pike. Outdoor Vehicle Navigation Using Passive 3D Vision. In: *Computer Vision and Pattern Recognition*, pp. 556–562. IEEE Computer Society Press, 1989.

3.19 T. Luong. *Matrice Fondamentale et Calibration Visuelle sur l'Environnement*. PhD thesis, Université de Paris-Sud, Orsay, 1992.

3.20 J.M. Lavest, G. Rives, and M. Dhome. 3D Reconstruction by Zooming. In: *Intelligent Autonomous System, Pittsburg*, 1993.

3.21 O.D. Faugeras. *Three-Dimensional Computer Vision: A Geometric Viewpoint*. MIT Press, Boston, 1993.

3.22 T. Viéville. Autocalibration of Visual Sensor Parameters on a Robotic Head. *Image and Vision Computing*, 12, 1994.

3.23 R. Lenz. *Group Theoretical Methods in Image Processing*. Lecture Notes in Computer Science 413, Springer-Verlag, Berlin, 1990.

3.24 N. Ayache. *Artificial Vision for Mobile Robots*. MIT Press, Cambridge, MA, 1989.

3.25 P.E. Gill and W. Murray. Algorithms for the Solution of Non-linear Least Squares Problem. *SIAM J. on Numerical Analysis*, 15:977–992, 1978.

3.26 T. Viéville and P. Sander. Using Pseudo Kalman-Filters in the Presence of Constraints. Technical Report RR-1669, INRIA, Sophia, France, 1992.

3.27 H.P. Trivedi. Estimation of Stereo and Motion Parameters Using a Variational Principle. *Image and Vision Computing*, 5, May 1987.

3.28 J.L. Crowley, P. Bobet, and C. Schmid. Autocalibration by Direct Observations of Objects. *Image and Vision Computing*, 11, 1993.

3.29 H.C. Longuet-Higgins. A Computer Algorithm for Reconstructing a Scene from Two Projections. *Nature*, 293:133–135, 1981.

3.30 S.J. Maybank. Properties of Essential Matrices. *International J. of Imaging Systems and Technology*, 2, 1990.

Chapter 4

4.1 P. Faurre. *Navigation Inertielle Optimale et Filtrage Statistique*. Dunod, Paris, 1971.

4.2 N. Ayache. *Artificial Vision for Mobile Robots*. MIT Press, Cambridge, MA, 1989.

4.3 J. Stuelpnagel. On the Parametrization of the Three-Dimensional Rotation Group. *SIAM Review*, 6:422–430, 1964.

4.4 M. Le Borgne. Quaternions et Contrôle sur l'Espace des Rotations. Technical Report 751, Institut National en Informatique et Automatique, Rocquencourt, 1987.

4.5 T. Viéville. Construction d'un Modèle Robotique du Contrôle Neurophysiologique des Mouvements Oculaires en Vue de l'Élaboration de Robots de 3ème Génération. Technical Report No. 86J0326, Ministère de la Recherche et de la Technologie, 1987.

4.6 T. Viéville and O.D. Faugeras. Cooperation of the Inertial and Visual Systems. In: *Traditional and Non-Traditional Robotic Sensors*, ed. by T. Henderson, pp. 339–350. Springer-Verlag, Berlin, September 1989.

4.7 R. Lenz. *Group Theoretical Methods in Image Processing*. Lecture Notes in Computer Science 413, Springer-Verlag, Berlin, 1990.

4.8 S. Samson, M. le Borgne, and B. Espiau. *Robot Control: The Task-Function Approach*. Oxford University Press, 1990.

4.9 T. Viéville, P.E.D.S. Facao, and E. Clergue. Computation of Ego-Motion Using the Vertical Cue. *Machine Vision and Applications*, 1994. To appear.

4.10 O.D. Faugeras. *Three-Dimensional Computer Vision: A Geometric Viewpoint*. MIT Press, Boston, 1993.

4.11 H. Cartan. *Cours de Calcul Différentiel*, Chapter: Equations Différentielles Linéaires, Chap. II-2. Hermann, Paris, 1982.

4.12 G. Hochschild. *La Structure des Groupes de Lie*. Dunod, Paris, 1968.

4.13 R.T. Collins and R.S. Weiss. Vanishing Point Calculation as a Statistical Inference on the Unit Sphere. In: *Proc. of the 3rd International Conference on Computer Vision, Osaka*, pp. 400–405. IEEE Computer Society Press, Alamitos, CA, 1990.

4.14 J.J. Gibson. *The Ecological Approach to Visual Perception*. Lawrence Erlbaum Associates, London, 1979.

4.15 S. Carlsson. Projectively Invariant Representation of Planar Shape. In: *5th Scandinavian Workshop on Computational Vision*, Royal College of Stockholm, Rosenon, Sweden, 1991.

4.16 R. Deriche and O.D. Faugeras. Tracking Line Segments. In: *Proc. of the 1st European Conference on Computer Vision, Antibes*, pp. 259–269. Springer-Verlag, Berlin, 1990.

4.17 T. Viéville and P. Sander. Using Pseudo Kalman-Filters in the Presence of Constraints. Technical Report RR-1669, INRIA, Sophia, France, 1992.

4.18 T. Viéville. Estimation of 3D-Motion and Structure from Tracking 2D-Lines in a Sequence of Images. In: *Proc. of the 1st European Conference on Computer Vision, Antibes*, pp. 281–292. Springer-Verlag, Berlin, 1990.

4.19 R. Deriche and B. Giai-Checa. Appariement de Segments dans une Sequence d'Images. In: *Orasis Meeting*, 1991.

4.20 T. Viéville and O.D. Faugeras. Feed Forward Recovery of Motion and Structure from a Sequence of 2D-Lines Matches. In: *3rd International Conference on Computer Vision, Osaka*, ed. by S. Tsuji, A. Kak, and J.-O. Eklundh, pp. 517–522. IEEE Computer Society Press, Los Alamitos, CA, 1990.

4.21 A. Mitiche, S. Seida, and J.K. Aggarwal. Interpretation of Structure and Motion Using Straight Line Correspondences. In: *Proc. of the 8th ICPR, Paris, France, October 1986*, pp. 1110–1112. IEEE Computer Society Press, Alamitos, CA.

4.22 Y. Liu and T.S. Huang. Estimation of Rigid Body Motion Using Straight Line Correspondences. *Computer Vision, Graphics and Image Processing*, pp. 35–57, 1988.

4.23 K. Kanatani. Hypothezing and Testing Geometric Attributes of Image Data. In: *Proc. of the 3rd International Conference on Computer Vision, Osaka*, 1990.

Chapter 5

5.1 E. Francois and P. Bouthemy. Multiframe-Based Identification of Mobile Components of a Scene with a Moving Camera. In: *Conference on Computer Vision and Pattern Recognition, Hawai*, pp. 166–172. IEEE Computer Society Press, Alamitos, CA, 1991.

5.2 N. Ayache and O.D. Faugeras. Determining Three-Dimensional Motion and Structure from Optical Flow Generated by Several Moving Objects. *IEEE Trans. on Pattern Analysis and Machine Intelligence*, 7:384–401, 1985.

5.3 S. Peleg and H. Rom. Motion Based Segmentation. In: *Proc. of the 10th IEEE Conference on Pattern Recognition, Atlantic City*, pp. 109–113, 1990.

5.4 D.W. Murray and H. Buxton. Scene Segmentation from Visual Motion Using Global Optimization. *IEEE Trans. on Pattern Analysis and Machine Intelligence*, 9:220–228, 1987.

5.5 O.D. Faugeras, Q.T. Luong, and S. Maybank. Camera Self-Calibration: Theory and Experiment. In: *2nd European Conference on Computer Vision*, Genoa, 1992.

5.6 H.P. Trivedi. Semi-analytic Method for Estimating Stereo Camera Geometry from Matched Points. *Image and Vision Computing*, 9, 1991.

5.7 N.A. Thacker. On-line Calibration of a 4-dof Robot Head for Stereo Vision. In: *British Machine Vision Association Meeting on Active Vision*, London, 1992.

5.8 T. Viéville. Autocalibration of Visual Sensor Parameters on a Robotic Head. *Image and Vision Computing*, 12, 1994.

5.9 M.J. Stephens, R.J. Blisset, D. Charnley, E.P. Sparks, and J.M. Pike. Outdoor Vehicle Navigation Using Passive 3D Vision. In: *Computer Vision and Pattern Recognition*, pp. 556–562. IEEE Computer Society Press, 1989.

5.10 R. Deriche and O.D. Faugeras. Tracking Line Segments. In: *Proc. of the 1st European Conference on Computer Vision, Antibes*, pp. 259–269. Springer-Verlag, Berlin, 1990.

5.11 R. Deriche and B. Giai-Checa. Appariement de Segments dans une Sequence d'Images. In: *Orasis Meeting*, 1991.

5.12 J.M. Lavest, G. Rives, and M. Dhome. 3D Reconstruction by Zooming. In: *Intelligent Autonomous System, Pittsburg*, 1993.

5.13 O.D. Faugeras. *Three-Dimensional Computer Vision: A Geometric Viewpoint*. MIT Press, Boston, 1993.

5.14 A.M. Waxman and S. Ullman. Surface Structure and Three-Dimensional Motion from Image Flow Kinematics. *International J. of Robot. Res.*, 4, 1985.

5.15 R. Tsai, T.S. Huang, and W.L. Zhu. Estimating Three-Dimensional Motion Parameters of a Rigid Planar Patch. ii: Singular Value Decomposition. *IEEE Trans. on Acoustic, Speech and Signal Processing*, 30:525–534, August 1982.

5.16 T. Viéville, C. Zeller, and L. Robert. Using Collineations to Compute Motion and Structure in an Uncalibrated Image Sequence. *International J. of Computer Vision*, 1995. To appear.

5.17 T. Viéville, P.E.D.S. Facao, and E. Clergue. Computation of Ego-Motion Using the Vertical Cue. *Machine Vision and Applications*, 1994. To appear.

5.18 Q.-T. Luong and T. Viéville. Canonic Representations for the Geometries of Multiple Projective Views. In: *3rd European Conference on Computer Vision, Stockholm*, 1994.

5.19 H.C. Longuet-Higgins. A Computer Algorithm for Reconstructing a Scene from Two Projections. *Nature*, 293:133–135, 1981.

5.20 T. Luong. *Matrice Fondamentale et Calibration Visuelle sur l'Environnement*. PhD thesis, Université de Paris-Sud, Orsay, 1992.

5.21 L. Robert and O.D. Faugeras. Relative 3D Positioning and 3D Convex Hull Computation from a Weakly Calibrated Stereo Pair. In: *4th International. Conference on Computer Vision, Berlin*, ed. by H.H. Nagel. IEEE Computer Society Press, Los Alamitos, CA, 1993.

5.22 Q.T. Luong and O.D. Faugeras. Determining the Fundamental Matrix with Planes: Instability and New Algorithms. In: *IEEE Proc. CVPR '93, New-York, June*, pp. 194–199, 1993.

5.23 O.D. Faugeras, F. Lustman, and G. Toscani. Motion and Structure from Point and Line Matches. In: *Proc. of the 1st International Conference on Computer Vision, London*, pp. 25–34, June 1987.

5.24 T. Viéville. Estimation of 3D-Motion and Structure from Tracking 2D-Lines in a Sequence of Images. In: *Proc. of the 1st European Conference on Computer Vision, Antibes*, pp. 281–292. Springer-Verlag, Berlin, 1990.

5.25 T. Viéville, J.O. Ekhlund, K. Pahlavan, and T. Uhlin. An Example of Artificial Oculomotor Behavior. In: *7th IEEE Symposium on Intelligent Control, Glasgow*, ed. by T. Henderson, pp. 348–353. IEEE Computer Society Press, 1992.

5.26 T. Viéville and P. Sander. Using Pseudo Kalman-Filters in the Presence of Constraints. Technical Report RR-1669, INRIA, Sophia, France, 1992.

5.27 T. Viéville, Q.T. Luong, and O.D. Faugeras. Motion of Points and Lines in the Uncalibrated Case. *International J. of Computer Vision*, 1994. To appear.

5.28 Q.T. Luong, R. Deriche, O.D. Faugeras, and T. Papadopoulo. On Determining the Fundamental Matrix: Analysis of Different Methods and Experimental Results. Technical Report RR-1894, INRIA, Sophia, France, 1993.

5.29 Y. Bar Shalom and T.E. Fortmann. *Tracking and Data Association.* Academic-Press, Boston, 1988.

5.30 P.R. Kumar and P. Varaiya, editors. *Stochastic Systems: Estimation, Identification and Adaptive Control.* Prentice Hall, New Jersey, 1986.

5.31 D.W. Murray, P.F. MacLauchlan, I.D. Reid, and P.M. Sharkey. Reactions to Peripheral Image Motion Using a Head/Eye Platform. In: *4th International Conference on Computer Vision*, pp. 403–411. IEEE Society, 1993.

5.32 P.E. Jupp and K.V. Martin. A Unified View of the Theory of Directional Statistics. *International Statistical Review*, 57, 1989.

5.33 R.T. Collins and R.S. Weiss. Vanishing Point Calculation as a Statistical Inference on the Unit Sphere. In: *Proc. of the 3rd International Conference on Computer Vision, Osaka*, pp. 400–405. IEEE Computer Society Press, Alamitos, CA, 1990.

5.34 B. Giai-Checa, P. Bouthemy, and T. Viéville. Detection of Moving Objects. Technical Report RR-1906, INRIA, Sophia, France, 1993.

5.35 O.D. Faugeras, B. Hotz, H. Mathieu, T. Viéville, Z. Zhang, P. Fua, E. Théron, L. Moll, G. Berry, J. Vuillemin, P. Bertin, and C. Proy. Real Time Correlation-Based Stereo: Algorithm, Implementations, and Applications. Technical Report 2013, INRIA, 1993.

5.36 M.J. Blake and P. Anandan. A Framework for the Robust Estimation of Optical Flow. In: *4th International Conference on Computer Vision*, pp. 231–236. IEEE Society, 1993.

5.37 Z. Zhang, R. Deriche, Q.T. Luong, and O. Faugeras. A Robust Approach to Image Matching: Recovery of the Epipolar Geometry. In: *Proc. of the International Symposium of Young Investigators on Information and Computer Control*, 1994. In press.

5.38 T. Viéville. Vision Modules for Active Vision. Technical Report, Université de Nice, 1993. Hand-Book of the Active Vision Course, INRIA, Sophia, France.

Chapter 6

6.1 H.C. Longuet-Higgins. A Computer Algorithm for Reconstructing a Scene from Two Projections. *Nature*, 293:133–135, 1981.

6.2 Y. Liu and T.S. Huang. Estimation of Rigid Body Motion Using Straight Line Correspondences. *Computer Vision, Graphics and Image Processing*, pp. 35–57, 1988.

6.3 A. Mitiche, S. Seida, and J.K. Aggarwal. Interpretation of Structure and Motion Using Straight Line Correspondences. In: *Proc. of the 8th ICPR*, pp. 1110–1112, Paris, France, October 1986. IEEE Computer Society Press, Alamitos, CA.

6.4 O.D. Faugeras, F. Lustman, and G. Toscani. Motion and Structure from Point and Line Matches. In: *Proc. of the 1st International Conference on Computer Vision, London*, pp. 25–34, June 1987.

6.5 O.D. Faugeras, Q.T. Luong, and S. Maybank. Camera Self-Calibration: Theory and Experiment. In: *2nd European Conference on Computer Vision*, Genoa, 1992.

6.6 H.P. Trivedi. Semi-analytic Method for Estimating Stereo Camera Geometry from Matched Points. *Image and Vision Computing*, 9, 1991.

6.7 N.A. Thacker. On-line Calibration of a 4-dof Robot Head for Stereo Vision. In: *British Machine Vision Association Meeting on Active Vision*, London, 1992.

6.8 T. Viéville. Autocalibration of Visual Sensor Parameters on a Robotic Head. *Image and Vision Computing*, 12, 1994.

6.9 R.I. Hartley and R. Gupta. Computing Matched-Epipolar Projections. In: *Proc. of the CVPR '93 Conference*, pp. 549–555, 1993.

6.10 R.I. Hartley. Camera Calibration Using Line Correspondences. In: *Proc. DARPA Image Understanding Workshop*, pp. 361–366, March 1993.

6.11 R. Deriche and O.D. Faugeras. Tracking Line Segments. In: *Proc. of the 1st European Conference on Computer Vision, Antibes*, pp. 259–269. Springer-Verlag, Berlin, 1990.

6.12 M.J. Stephens, R.J. Blisset, D. Charnley, E.P. Sparks, and J.M. Pike. Outdoor Vehicle Navigation Using Passive 3D Vision. In: *Computer Vision and Pattern Recognition*, pp. 556–562. IEEE Computer Society Press, 1989.

6.13 J.L. Crowley, P. Bobet, and C. Schmid. Autocalibration by Direct Observations of Objects. *Image and Vision Computing*, 11, 1993.

6.14 J.M. Lavest, G. Rives, and M. Dhome. 3D Reconstruction by Zooming. In: *Intelligent Autonomous System, Pittsburg*, 1993.

6.15 R.Y. Tsai. Synopsis of Recent Progress on Camera Calibration for 3D Machine Vision. *Robotics Review*, 1:147–159, 1989.

6.16 O.D. Faugeras. *Three-Dimensional Computer Vision: A Geometric Viewpoint*. MIT Press, Boston, 1993.

6.17 S. Maybank and O.D. Faugeras. A Theory of Self-Calibration of a Moving Camera. *The International J. of Computer Vision*, 8, 1992.

6.18 R. Enciso, T. Viéville, and O. Faugeras. Approximation du Changement de Focale et de Mise au Point par une Transformation Affine à Trois Paramètres. Technical Report 2071, INRIA, 1993.

6.19 R. Willson. *Modeling and Calibration of Automated Zoom Lenses*. PhD thesis, Department of Electrical and Computer Engineering, Carnegie Mellon University, 1994.

6.20 R.G. Willson and S.A. Shafer. What is the Center of the Image ? In: *IEEE Proc. CVPR '93, New York, June*, pp. 670–671, 1993.

6.21 T. Viéville, P.E.D.S. Facao, and E. Clergue. Building a Depth and Kinematic 3D-Map from Visual and Inertial Sensors Using the Vertical Cue. In: *4th International. Conference on Computer Vision, Berlin*, ed. by H.H. Nagel. IEEE Computer Society Press, Los Alamitos, CA, 1993.

6.22 T. Viéville, C. Zeller, and L. Robert. Using Collineations to Compute Motion and Structure in an Uncalibrated Image Sequence. *International J. of Computer Vision*, 1995. To appear.

6.23 T.S. Huang and A. Netravali. Linear and Polynomial Methods in Motion Estimation. In: *Signal Processing, Part I: Signal Processing Theory*, ed. by L. Auslander, T. Kailath, and S. Mitter. Springer-Verlag, 1990.

6.24 J.L.S. Mundy and A. Zisserman. *Geometric Invariance in Computer Vision*. MIT Press, Boston, 1992.

6.25 T. Luong. *Matrice Fondamentale et Calibration Visuelle sur l'Environnement*. PhD thesis, Université de Paris-Sud, Orsay, 1992.

6.26 Q.-T. Luong and T. Viéville. Canonic Representations for the Geometries of Multiple Projective Views. In: *3rd European Conference on Computer Vision, Stockholm*, 1994.

6.27 O.D. Faugeras. What Can Be Seen in Three Dimensions with an Uncalibrated Stereo Rig? In: *2nd European Conference on Computer Vision, Genoa*, 1992.

6.28 L Quan. Invariants of 6 Points form 3 Uncalibrated Images. In: *3rd European Conference on Computer Vision, Stockholm*, 1994.

6.29 N. Navab, O.D. Faugeras, and T. V iéville. The Critical Sets of Lines for Camera Displacement Estimation:A Mixed Euclidean-Projective and Constructive Approach. In: *Proc. of the 4th International Conference on Computer Vision, Berlin, Germany, May 1993*, pp. 713–723, IEEE.

6.30 T. Viéville and O.D. Faugeras. Feed Forward Recovery of Motion and Structure from a Sequence of 2D-Lines Matches. In: *3rd International Conference on Computer Vision, Osaka*, ed. by S. Tsuji, A. Kak, and J.-O. Eklundh, pp. 517–522. IEEE Computer Society Press, Los Alamitos, CA, 1990.

6.31 L. Robert. *Perception Stéréoscopique de Courbes et de Surfaces Tridimensionnelles, Application à la Robotique Mobile*. PhD thesis, Ecole Polytechnique, Palaiseau. France, 1992.

6.32 L. Robert and O.D. Faugeras. Relative 3D Positioning and 3D Convex Hull Computation from a Weakly Calibrated Stereo Pair. In: *4th International. Conference on Computer Vision, Berlin*, ed. by H.H. Nagel. IEEE Computer Society Press, Los Alamitos, CA, 1993.

6.33 Q.T. Luong, R. Deriche, O.D. Faugeras, and T. Papadopoulo. On Determining the Fundamental Matrix: Analysis of Different Methods and Experimental Results. Technical Report RR-1894, INRIA, Sophia, France, 1993.

6.34 J. Heel. Temporally Integrated Surface Reconstruction. In: *Proc. of the 3rd International Conference on Computer Vision, Osaka*, 1990.

6.35 T. Viéville, P.E.D.S. Facao, and E. Clergue. Computation of Ego-Motion Using the Vertical Cue. *Machine Vision and Applications*, 1994. To appear.

6.36 P.E. Gill and W. Murray. Algorithms for the Solution of Non-linear Least Squares Problem. *SIAM J. on Numerical Analysis*, 15:977–992, 1978.

6.37 R. Deriche and G. Giraudon. Accurate Corner Detection: An Analytical Study. In: *Proc. of the 3rd International Conference on Computer Vision, Osaka*, pp. 66–71, 1990.

6.38 A. Guiducci. Corner Characterization by Differential Geometry Techniques. *Pattern Recognition Letters*, 8:311–318, 1988.

6.39 P.A. Ruymgaart and T.T. Soong. *Mathematics of Kalman-Bucy filtering*. Springer-Verlag, Berlin, 1985.

6.40 L.N. Kanal and J.F. Lemmer. *Uncertainty in Artificial Intelligence*. North-Holland Press, Amsterdam, 1988.

Chapter 7

7.1 D.W. Murray, P.F. MacLauchlan, I.D. Reid, and P.M. Sharkey. Reactions to Peripheral Image Motion Using a Head/Eye Platform. In: *4th International Conference on Computer Vision*, pp. 403–411. IEEE Society, 1993.

7.2 E. Francois and P. Bouthemy. Derivation of Qualitative Information in Motion Analysis. *Image and Vision Computing*, 8, 1990.

7.3 R.J. Leight and D.S. Zee. *The Neurophysiology of Eye Movement*. Davis Cie, Philadelphia, 1984.

7.4 A. Berthoz and G.M. Jones. *Adaptive Mechanism in Gaze Control*. Elsevier, Amsterdam, 1985.

7.5 T. Viéville, S. Ron, and J. Droulez. Two Dimensional Saccadic and Smooth Pursuit Response to an Extrafoveal Smooth Movement. In: *Proc. of the 3rd European Conference on Eye Movements*, ed. by A. Levy-Shoen and K. O'Reagan, 1986.

7.6   K. Pahlavan, T. Uhlin, and J.-O. Ekhlund. Dynamic Fixation. In: *4th International Conference on Computer Vision*, pp. 412–419. IEEE Society, 1993.

7.7   N.P. Papanikolopoulos, B. Nelson, and P.K. Khosla. Full 3D Tracking Using the Controlled Active Vision Paradigm. In: *The 7th IEEE Symposium on Intelligent Control, Glasgow, August*, 1992.

7.8   E. Francois and P. Bouthemy. Multiframe-Based Identification of Mobile Components of a Scene with a Moving Camera. In: *Conference on Computer Vision and Pattern Recognition, Hawai*, pp. 166–172. IEEE Computer Society Press, Alamitos, CA, 1991.

7.9   R. Enciso, T. Viéville, and O. Faugeras. Approximation du Changement de Focale et de Mise au Point par une Transformation Affine à Trois Paramètres. Technical Report 2071, INRIA, 1993.

7.10 R. Willson. *Modeling and Calibration of Automated Zoom Lenses*. PhD thesis, Department of Electrical and Computer Engineering, Carnegie Mellon University, 1994.

7.11 K. Pahlavan, J.-O. Ekhlund, and T. Uhlin. Integrating Primary Ocular Processes. In: *2nd European Conference on Computer Vision*, pp. 526–541. Springer-Verlag, 1992.

7.12 K. Pahlavan, J.-O. Ekhlund, and T. Uhlin. Calibration of the *k*th Head. In: *3rd European Conference on Computer Vision*, 1994. In Press.

7.13 O.D. Faugeras, Q.T. Luong, and S. Maybank. Camera Self-Calibration: Theory and Experiment. In: *2nd European Conference on Computer Vision*, Genoa, 1992.

7.14 H.P. Trivedi. Semi-analytic Method for Estimating Stereo Camera Geometry from Matched Points. *Image and Vision Computing*, 9, 1991.

7.15 N.A. Thacker. On-line Calibration of a 4-dof Robot Head for Stereo Vision. In: *British Machine Vision Association Meeting on Active Vision*, London, 1992.

7.16 Q.T. Luong, R. Deriche, O.D. Faugeras, and T. Papadopoulo. On Determining the Fundamental Matrix: Analysis of Different Methods and Experimental Results. Technical Report RR-1894, INRIA, Sophia, France, 1993.

7.17 Y. Aloimonos, I. Weiss, and A. Bandopadhay. Active Vision. *International J. Comp. Vision*, 7:333–356, 1988.

7.18 R. Bajcsy. Active Perception. *Proc IEEE 76*, 8:996–1005, 1988.

7.19 D. Ballard. Animate Vision. *Artificial Intelligence*, 48:57–86, 1991.

7.20 J.J. Gibson. *The Ecological Approach to Visual Perception*. Lawrence Erlbaum Associates, London, 1979.

7.21 I.P. Howard. *Human Visual Orientation*. John Wiley and Sons, New York, 1982.

7.22 T. Viéville and O.D. Faugeras. Computation of Inertial Information on a Robot. In: *5th International Symposium on Robotics Research*, ed. by H. Miura and S. Arimoto, pp. 57–65. MIT-Press, 1989.

7.23 O.D. Faugeras and T. Viéville. Cooperation of the Inertial and Visual Sensors. In: *International Workshop on Sensorial Integration for Industrial Robots: Architectures and Applications, Zaragoza, Spain*, ed. by T. Lozano-Perez. November 1989.

7.24 T. Viéville and O.D. Faugeras. Cooperation of the Inertial and Visual Systems. In: *Traditional and Non-Traditional Robotic Sensors*, ed. by T. Henderson, pp. 339–350. Springer-Verlag, Berlin, September 1989.

7.25 T. Viéville, P.E.D.S. Facao, and E. Clergue. Building a Depth and Kinematic 3D-Map from Visual and Inertial Sensors Using the Vertical Cue. In: *4th International. Conference on Computer Vision, Berlin*, ed. by H.H. Nagel. IEEE Computer Society Press, Los Alamitos, CA, 1993.

7.26 T. Viéville, F. Romann, B. Hotz, H. Mathieu, M. Buffa, L. Robert, P.E.D.S. Facao, O.D. Faugeras, and J.T. Audren. Autonomous Navigation of a Mobile Robot Using Inertial and Visual Cues. In: *Intelligent Robots and Systems*, ed. by M. Kikode, T. Sato, and K. Tatsuno. Yokohama, 1993.

7.27 T. Viéville, Q.T. Luong, and O.D. Faugeras. Motion of Points and Lines in the Uncalibrated Case. *International J. of Computer Vision*, 1994. To appear.

7.28 Q.-T. Luong and T. Viéville. Canonic Representations for the Geometries of Multiple Projective Views. In: *3rd European Conference on Computer Vision, Stockholm*, 1994.

7.29 T. Viéville, B. Hotz, H. Mathieu, M. Buffa, L. Robert, L. Robert, P.E.D.S. Facao, O.D. Faugeras, F. Romann, and J.T. Audren. Using Visual Feedback to Drive a Mobile Robot on its Trajectoy. In: *Workshop on Computer Vision for Space Applications*, ed. by G. Giraudon and M. Plancke, 1993.

Index

240   Index

Springer Series in Information Sciences

Editors: Thomas S. Huang   Teuvo Kohonen   Manfred R. Schroeder

Managing Editor: H. K. V. Lotsch

Springer
and the
environment

At Springer we firmly believe that an international science publisher has a special obligation to the environment, and our corporate policies consistently reflect this conviction.

We also expect our business partners – paper mills, printers, packaging manufacturers, etc. – to commit themselves to using materials and production processes that do not harm the environment. The paper in this book is made from low- or no-chlorine pulp and is acid free, in conformance with international standards for paper permanency.